MASSEY
AT THE BRINK

MASSEY AT THE BRINK

The story of Canada's
greatest multinational
and its struggle
to survive

PETER COOK

Collins Toronto

To Claudia and Tasha Jane

First published 1981
by Collins Publishers
100 Lesmill Road, Don Mills, Ontario

© 1981 by Peter Cook

Canadian Cataloguing in Publication Data

Cook, Peter, 1941-
    Massey at the brink

Includes index.
ISBN 0-00-216857-X

1. Massey Ferguson Ltd. — Finance — History.
2. Massey Ferguson Ltd. — History. I. Title.

HD9486.C24M38   338.7′681763′0971   C81-094987-3

Printed and bound in Canada by
T. H. Best Printing Company Limited, Don Mills, Ontario

Contents

Part Three:
The Thornbrough Years (1956-1978)

Part Four:
At the Brink (1978-1981)

Epilogue

Acknowledgements

I would like to thank the people of Massey, past and present, who helped me in small and large ways with the preparation of this book. Many of them agreed to be interviewed, often at great length, with no guarantees about how the company, or their role in it, would be portrayed. Others helped in other ways. They include: Michael Bird, Peder Bjerre, Roger Clarke, Bob Doll, Glenn Frederick, Darwin Kettering, Rik Kettle, Vincent Laurenzo, Brian Long, Peter Lowry, Philip Moate, Victor Rice, John Ruth, John Staiger, Albert Thornbrough, Ebbie Whittington. My special thanks go to Peter Lowry for his invaluable help.

I also owe a debt to a number of authors who have written about the company, about the industry and about the Canadian business scene. Among the sources which I have drawn upon and which have proved most informative are:

Paul Collins *Hart Massey*, Toronto: Fitzhenry and Whiteside, 1977.

Mollie Gillen *The Masseys: Founding Family*. Toronto: Ryerson Press, 1965.

Raymond Massey *When I was young*. Toronto: McClelland and Stewart, 1976.

Vincent Massey *What's Past is Prologue*. Toronto: Macmillan Co. of Canada Ltd, 1963.

James S. Duncan *Not a One-Way Street*. Toronto: Clarke Irwin, 1971.

Royal Commission on Corporate Concentration "Study No.1, Argus Corp. Ltd: a Background Report." H.T. Seymour, Pitfield Mackay Ross and Co. Ltd. Ottawa: Supply and Services Canada, 1977.

E.P. Neufeld *A Global Corporation*. Toronto: University of Toronto Press, 1969.

Richard Rohmer *E.P. Taylor*. Toronto: McClelland and
    Stewart, 1978.
Colin Fraser *Harry Ferguson: Inventor and Pioneer*. London:
    John Murray, 1972.
George Bookman "Farm Machinery Shifts Gear." *Fortune
    Magazine*, July, 1961.
William B. Harris "Massey Ferguson: Back from the Brink."
    Fortune Magazine, October, 1958.
Graham Turner *The Leyland Papers*. London: Eyre and
    Spottiswoode, 1971.
Peter C. Newman *The Canadian Establishment*. Toronto:
    McClelland and Stewart, 1975.

Finally I would like to thank my friend and former colleague,
Terry Corcoran, and my editor at Collins, Margaret Paull, for
their support and advice with this project.

Peter
Cook

1
A Company
Worth Saving

*As the bankers and lawyers perspired, the final
ritual was played out in Massey's long struggle
to avoid financial disaster and bankruptcy.*

On a hot July morning in 1981, a group of international
bankers filed into a nondescript reception room on the mezzanine
floor of Toronto's Royal York Hotel. Once inside the room,
they were greeted by the president and chairman of Massey-
Ferguson, Victor Rice, and took their places around a long
rectangular table. There were cards denoting where the bankers,
from Britain, Germany, the United States, and Canada,
were to sit but the crush of bankers (who were accompanied
by their lawyers and tax accountants, and a bevy of Canadian
government officials), was too much for the small room. Many
had to stand or sit at the back next to a table of telephones and
telex machines marked with the names of the cities to which
calls would be placed — Paris, London, Zurich. To add to the
discomfort television lights were switched on and cameras began
to film the proceedings.

During the next hour, as the bankers and lawyers perspired,
the final ritual was played out in Massey's long struggle to avoid
financial disaster and bankruptcy. One by one, the bankers
were called upon to table thick loan documents and certificates,
and to acknowledge that they would go along with a refinancing
scheme that would forgive interest payments and give the
company a chance to survive. At the head of the table, the
chairman of the meeting, Bill Corbett, representing Massey's
legal firm Fraser and Beatty, worked methodically through an
eighteen-point agenda. First the banks had to agree to the terms;

10 then the most crucial part of the refinancing had to go smoothly, the issue of preferred shares that would give Massey a life-saving infusion of new equity. Guaranteed by the federal government in Ottawa and the provincial government of Ontario, the new shares and warrants would provide $200 million in cash, a sum desperately needed to keep the company afloat.

The cheques and bankers' drafts that were put on the table were for huge amounts; Massey's own bank, the Canadian Imperial Bank of Commerce (CIBC), represented by a senior executive, Tom Grindly, handed over a draft for $70 million; John McMillan, of the Royal Bank of Canada, delivered a cheque for $100 million for the preferred shares and several million more for the purchase of warrants attached to the shares. Then the process of telexing and telephoning banks and financial institutions around the world began. In other countries, bankers were waiting for word that the final deal had been concluded. Only when they knew that things had gone smoothly in Toronto would they release their own loan agreements, and allow the transfusion of funds from one subsidiary company in Massey's worldwide network to another to occur.

The closing agreement had been prepared in advance. For three days, Massey's lenders had been checking documents and shuffling between their hotel and the company's head office to satisfy themselves that their own bank — and the others — were getting a fair deal. Still, there had been last-minute upsets. Three U.S. banks, accounting for $27 million out of a $380 million loan syndication to Massey in the U.S., balked at signing. The three banks, the National Bank of Detroit, the First Tennessee Bank, and the First National Bank of Maryland, did not want to join a scheme that would forgive interest on Massey's debt and preferred to be paid off.* Since the defection of even one would have put the whole refinancing in jeopardy, they had to be won over.

By the final morning, all the obstacles had been overcome. Among those gathered around the table, the most important participants were representatives of the two leading Canadian banks, the CIBC and the Royal; two U.S. banks, the Continental Illinois National Bank of Chicago and Citibank of New York; and Britain's Barclays Bank.

As the meeting progressed, there was an anxious wait for a telex message from Paris. A banker named Pichot of Société Générale had told Massey officials that he would be at an

---

* One of the U.S. bankers who objected startled his colleagues at an early meeting of lenders by announcing that his bank intended to play "hard ball" with Massey all the way.

important meeting and could not be disturbed. The French had signed a protocol ahead of time subject to agreement by the other banks; all that was required was for them to confirm their participation, and this Pichot undertook to do by telex. When the telex arrived in the offices of Fraser and Beatty, there was no one there who read French and could translate it. As 300 people waited in the hotel room, the telex was read over letter by letter to a French-speaking Massey executive, Brian Long, who was able to report back that the French, like all the others, would go along. Other snags that occurred were more in the nature of light relief than serious holdups; one Canadian trust company had failed to get its cheque for several million dollars certified and Rice was asked whether he would accept it, which he did. Another trust company, Canada Trust, presented a fully certified cheque for $10 million drawn on its own "corpaorate [sic]" loans account.

The closing of the Massey refinancing involved altogether a handout of $715 million. There was a sense of history as well as drama about the occasion. Rice duly noted that the proceedings had taken exactly fifty-four minutes, and the meeting was terminated by a Massey lawyer theatrically waving a red hotel napkin at the far end of the room to signify that the final calls, to the French, had been put through. At that point the lenders and their advisers were invited to a more spacious suite on the eighteenth floor of the hotel. Ten months earlier, setting out on the road to get his financing agreement, Rice had promised the Canadian press corps that he would drink champagne with them to celebrate its completion. The journalists were waiting upstairs. Rice swept in, waving the cheques that he had scooped up downstairs, and champagne corks were popped as he posed for photographers. Later in the day, he fulfilled another commitment; the staff at Massey's headquarters celebrated the survival of their company — and their jobs — over lunch and more champagne.

The congratulations and the sense of victory were deserved. Over a long and harrowing period of time, Massey had managed to patch together a formula for saving itself that was extraordinarily complicated while at the same time based on a simple premise. That premise, which Rice had managed to sell to hard-bitten financial men and politicians as well as to his own corporate team, was that Massey-Ferguson was worth saving. The company had contributed a great deal during its 140 years of history. And it must not be allowed to perish.

In the complex world of multinational business, there have been many instances when companies, finding themselves in trouble, have felt it necessary to appeal to their creditors or

12   to governments to prolong their life, holding off on repaying their debts or obtaining guarantees for new loans. The frequency with which companies have had to resort to such measures has, if anything, increased. In the inflationary environment of the late 1970s and early 1980s, there are plenty of examples of companies that have misread the signals, overcommitted themselves, and then found it impossible to retrench fast enough. For smaller businesses, the outcome has invariably been bankruptcy. For the major multinationals, banks and governments have found it harder to be tough, to countenance a corporate failure that would create the loss of thousands of jobs and the closure of industrial plants around the world. Instead, they have been prepared to go to extraordinary lengths to ensure that the industrial giant survives, even if it does so for only a limited period of time and in a greatly altered form.

The story of Massey-Ferguson, the huge Canadian farm equipment company, is the story of one such rescue bid. For three years, as it reported heavy losses and its management tried to cope with one crisis after another, Massey had been the subject of speculation in the financial community and on the financial pages of newspapers around the world. The stakes were high. Massey and its subsidiaries controlled a global empire that Rice had once described as being as widespread and all-enveloping as Coca-Cola. In many of the poorer countries of the world, Massey machinery was contributing mightily to the basic business of providing farm products and food. At the same time, the company had concentrated its investment, and the rich pool of jobs that it provided in the world's wealthy countries, and it was to the banks and the governments in these nations that the company appealed for help. With so much at stake, $3 billion in annual sales and the jobs of 45,000 employees around the world. Massey was a prime candidate for assistance. New life had to be breathed back into this company, or, at least, the attempt had to be made.

The deal was possibly the most convoluted and voluminous in corporate history.

The exact point at which Massey's problems became overwhelming can be traced back to 1978 when its losses began to pile up, and new management was put in place to cut and pare away and try to turn things around. In the two-year period

that followed, outside events conspired against the company;
farm markets did not recover as expected from their downturn
in the wake of the energy crisis, while rising interest rates pushed
the company into a position of default on interest payments
and debt covenants. By September, 1980, an urgent situation
had become critical. The banks who were Massey's principal
lenders could have forced the issue and precipitated a bankruptcy; the company had conceded that it did not have
sufficient cash to meet payments on its maturing debt or to
meet interest costs. To avoid an actual failure, the Massey
management canvassed the idea of a corporate refinancing. If
the lenders would hold off, and governments would help
sponsor a new infusion of equity, the worst might be prevented
from happening and Massey would be able to continue in
business: It would have some chance of survival.

In many respects, the refinancing plan was unique. Moreover, it was to prove uniquely difficult to put into effect since
it was not until ten months later that representatives from the
banks and financial institutions gathered in Toronto to give
final approval to their own part in the rescue.

The deal was possibly the most convoluted and voluminous
in corporate history. Mountains of documents and legal opinions
were involved in an attempt to strike a balance between
the competing claims of financial institutions in eight countries,
Canada, the United States, Britain, France, West Germany, the
Netherlands, Italy, and Argentina. Absent from the final
negotiations were lenders from Brazil and Australia who preferred to strike a deal on their own.

What the documents essentially did was give Massey time
to recoup. The lenders would undertake to forgive interest, to
the extent of 22.5 percent for senior banks and 30 percent for
surbordinated lenders, and restructure the company's huge debts
so that maturities would be deferred. Massey's leading banker,
the CIBC, agreed to convert a portion of debt into equity and to
inject new cash into the firm. The banks, however, were not
acting out of sentiment but out of hard self-interest. Each of the
more than 200 lenders was prepared to go along with a world-wide
interest forgiveness plan only on the understanding that its own
interests were being safeguarded. All of them wanted to build into
the agreement restrictive financial convenants, and all wanted to
avoid a situation in which assets would start moving from one
susidiary to another within the company, and from one national
jurisdiction to another which might be out of reach.

Essentially, Massey had three main subsidiaries that
generated most of the cash for the firm and were the focus of
all the attention: its British subsidiary, Massey-Ferguson (United

14   Kingdom) Limited; its U.S. operating company, Massey-Ferguson Incorporated, registered in the State of Delaware; and its Canadian operating company, Massey-Ferguson Industries Limited. As the bankers saw it, their duty to their shareholders required them to extract the maximum number of concessions and to tie Massey's future to its ability to overcome its indebtedness and repay its massive loans.

In its parlous state, Massey could scarcely afford to negotiate its way out of these concessions. Left to its own devices, the company might well have found itself tied down and prevented from resurrecting itself because of the onerous terms set by its lenders. There were precedents for this impasse in the rescue proposals advanced to bail out other farm machinery companies such as International Harvester and the White Motor Corporation in the U.S. Both had been so restricted by banks' demands for collateral, that the attempted rescue had been undermined. Harvester had managed to limp along, but in the winter of 1981 White had filed for court protection under the U.S. Bankruptcy Act.

The saving grace for Massey was the presence of two levels of Canadian government. Both the federal government and the provincial government of Ontario had agreed to guarantee the capital risk on a $200 million issue of preferred shares. The condition was that there must be a satisfactory amount of assistance from the lenders, and the banks and other financial institutions must be willing to take an additional risk, through acquiring warrants to purchase Massey common shares. If any dividend on the preferred shares were missed then the governments would agree to buy the shares at their original issue price.

The terms for the involvement of the governments gave Massey a trump card to play against the banks. It meant that there would be no government participation, and no extra infusion of needed equity, unless the banks held off and watered down their demands. Massey was a better prospect with this additional aid than without it. And the governments had a vested interest in keeping the company going thereby avoiding having to trigger their commitment. Massey must not be put in the position where it failed to pay a dividend on the new preferred shares, and the governments would do all in their power to ensure that it would not be.

Helpful in itself, this governmental co-operation had been hard-won and reluctantly given. In theory, Massey had enlisted the aid of the politicians and bureaucrats to do battle on its behalf. In practice, the support it received was never more than lukewarm. In addition, in the case of Ottawa, this support was handled in a manner that greatly added to the complication of what was being attempted.

In Ontario, the support for Massey took a clear and definite legislative form, but the Trudeau government chose not to face Parliament with the issue of its support for a major multinational corporation. Instead, it decided to bypass any debate on the matter and rely on existing powers to provide a guarantee in the event of Massey's failure to meet its commitments. It was a typically devious manoeuvre, and one that was to enrich an army of lawyers who had to consider the extent to which Ottawa's commitment would be legally binding on the government and therefore acceptable to the new shareholders.*

Inside Massey, there were those who noted rather sourly the rise and fall in government enthusiasm. One executive held that on a scale of one to ten (with five being taken as "entirely neutral"), Ottawa had initially responded the previous September at six, while the government of Ontario had been at five. Six months later, as things came to the crunch, Ontario had advanced to nine while the Trudeau government remained at seven. The issue of aid to Massey had split the federal cabinet. Still, a decision had been taken in favor of supporting the company and under the rules of cabinet solidarity, this meant the government was committed.

The scale of the refinancing, the prodigious effort required to get it under way, and the perseverance displayed by Massey's management were in themselves remarkable. They were in line with the seriousness of Massey's problems and the traditions and loyalties that surrounded the firm. The Canadian bankers who came to the rescue were aware that they were stepping in to save "a grand old Canadian company." The Massey management, pragmatic throughout, were obsessed not only with the idea that their company (the oldest and largest industrial enterprise the country had ever produced) was worth saving but that, after years of lurching from one crisis to another, anything was worth trying.

The attitude was summed up by one executive, Phil Moate, who declared that after "going through each day with the knowledge that we might be shot at dawn, all of us developed a real appreciation for what had to be done."

---

* The drawn-out negotiations were vexing and time-consuming for all the participants, and hugely rewarding for the legal advisers. Lawyers had worked on the Massey documents and had been called on to give their opinions in all the financial centres where the company had borrowed extensively; New York, Chicago, London, Paris, and Frankfurt. It was Massey's own lawyers, Fraser and Beatty in Toronto and Cahill Gordon in New York who pocketed the richest fees. Fraser and Beatty had been called on to give legal opinions on every single loan agreement, a monumentally rewarding task that netted the firm record fees. In all, over $15 million were passed out to the various law firms.

16   The sense of commitment, and of the need to scramble to survive, had been made necessary over several years. In the period leading up to the refinancing Massey had been through a time of utter turmoil, of management changes, of losses and retrenchment and cutbacks, and of internal fights and bickering. It had also come to earn for itself the unenviable reputation of being seen as a classic corporate failure, a poorly managed old-line manufacturing firm that had been unable to move with the times and had consistently made the wrong decisions. How much of this reputation was deserved was a matter of debate. But the company had hit the headlines as frequently as another industrial giant, Chrysler, and though Victor Rice hated the idea of Massey being bailed out by government, it was Canadian taxpayers who had in the end underwritten the company's future and insured that it would survive.

In Canada and in a number of other countries including Britain, the U.S., France, Germany, Italy, Brazil, Australia, and Argentina, Massey had a well-established name and reputation. But, historically, its name was most closely linked to the business community in Canada, its home base. Since the beginnings of farm mechanization and its founding as the Massey Manufacturing Company in the 1840s, the company had been a corporate blue-blood linked to the great names of Canadian business — the Massey family, the Gundys, James Duncan, E.P. Taylor, Eric Phillips, Bud McDougald, and young Conrad Black.

Massey had enjoyed periods of tremendous success, of innovation and expansion, and it had gone through other periods when through poor luck or incompetent management it had come close to failing altogether. If there was a common thread running through the history of the company, it was the tremendous influence exerted by its chief executives, from Daniel Massey to Victor Rice, and the regularity with which — generation after generation — they would find themselves battling against the odds and striving to keep Massey intact and solvent. From its base in rural Ontario, the company had expanded globally, achieving a position as the most important (though not the largest) maker of farm machinery in the world. Yet it was always vulnerable, and always as close to failure as to success.

Founded by Daniel Massey, the company under his stern and ambitious son, Hart, had swallowed up other Canadian firms to make itself the biggest farm equipment business of its time. In the 1920s, Hart's grandson Vincent had sold out, intent on pursuing a diplomatic career that was to lead to his appointment as the first Canadian-born Governor-General.

Massey was beset by crises during the 1920s and later

during the Depression. The company was only able to fully surmount their difficulties when, during the Second World War, it turned to producing armaments for the Canadian, British, and U.S. governments.

By the end of the Depression, Massey had found a savior in James Duncan. Over the next twenty years, Duncan displayed great tenacity and personal drive in remaking Massey into a great international company, building on the strengths the company had acquired during the war years. But he made a single tactical error. Inviting two wartime colleagues, E.P. Taylor and Eric Phillips, onto the Massey board of directors, he put them into a position of power where they could use their money and their influence against him. By the 1950s, Massey was in trouble again and a struggle for control ensued in which Duncan, chairman, president, and chief executive, was unceremoniously removed from his post. Eric Phillips took over, promoting an American, Albert Thornbrough, to president.

The 1960s and early 1970s were generally good years for Massey, as the company benefited from the internationalism of its new owners, Phillips, Taylor, and Bud McDougald, the powerful triumverate who ran Argus Corporation. Towards the end of the 1970s, the reins of power at Argus and at Massey had passed to McDougald alone. Through Argus, McDougald held an unrivalled position as Canada's most powerful businessman, ruling an empire that consisted of some of the country's biggest companies in retail food, broadcasting, mining, and forest products. For McDougald, and for Argus, Massey-Ferguson was the most profitable and best-known company in the Argus portfolio. In 1976, Massey reported record sales and record profits, and McDougald had congratulated its president, Albert Thornbrough, on its immense success. As it turned out, that success was to be both shallow and insubstantial.

McDougald died at his Florida home in March, 1978 and a fierce struggle began to rage for control of Argus, and the inheritance that went with it. At the same time Albert Thornbrough was having to tell Massey shareholders that the company would suffer huge losses. Over the next few months it became evident that disastrous management decisions, and Massey's beleaguered financial state, had put the survival of the company in doubt.

By the summer of 1978, financier Conrad Black had shouldered other contenders aside to seize power at Argus and install a new president, Victor Rice, to run Massey. For the next three years, the financial problems and pressures were to mount, culminating in Black's decision to give away the Argus share in Massey, and in a desperate bid by Rice to win a reprieve from

18 bankers and governments to avert the company's slide into bankruptcy.

The latest crisis remains the most serious in the company's history, so serious that Massey's survival is still in doubt. But there have been crises before, whether brought on by external events or internal divisions and misjudgments. At any time in the past three years as Victor Rice contemplated the chances of Massey winning through, he could have summoned up many parallel situations from the past. For one of Massey's most striking characteristics, stretching back to the 1840s, has been its resilience and ability to survive adversity.

2

The Glorious Age of the Machine

American companies found it simpler to sell their designs to Massey and Harris than to attempt to wrest their markets away from them. Transportation between the two countries was still poor. Rather than sell to Canadians directly, the Americans stayed at home.

In the history of any family, there are many differing influences at work. But for the Massey family — who were to become the founders of a great industrial empire — there were two great formative influences which seemed to drive them towards accomplishment. The first was the fervor of the family's nonconformist past which had made the Masseys of the seventeenth century uproot themselves from the English county of Cheshire and emigrate to Massachusetts. The second was the pioneer spirit of the New World which sent successive generations away from the comparative security of Salem, Massachusetts, into the wilderness areas of the Adirondacks and western New York state.

Geoffrey and Ellen Massey were the original settlers in the New World. By the time of the revolutionary war, their direct descendant Jonathan Massey had moved to New Hampshire and was commissioned into George Washington's army as a lieutenant. He and his wife Betsey had twelve children. Their eldest son Daniel and another son Hart travelled widely, journeying north towards the St. Lawrence River. Together, they built a log cabin in the back country and began to farm the surrounding lands. Their travels had taken them close to the small community of Watertown, New York, and it was

20 here that Hart decided to settle down. The two of them had by then been joined by their father. In the years that followed, Hart was to become a respected member of the community and a co-founder of the new city of Watertown. The city, close to the Canadian border, was to be the family home of the Masseys for this and later generations. And the Masseys were to remain a family with strong roots in the United States as well as in Canada.

Daniel, however, did not stay with the others. He had heard reports of the rich farmlands on the north shore of Lake Ontario which had attracted loyalist settlers and immigrants from the British Isles. With his wife Rebecca and three young sons — including their youngest who was also called Daniel — they sailed from Sackett's Harbor to the Bay of Quinte on the Canadian side of the lake. There, they rented and cleared land and began farming close to the settlement of Grafton.

Six years after they arrived in Upper Canada, the census rolls depict the Massey family as it then was. There were now three small daughters, making six children in all. Daniel Massey held one hundred acres of land. Seventy acres were forested, while thirty acres had been cleared for cultivation. The family had three horses, four oxen, and two cows. The total value of their small holding was £82.

It was in these surroundings that their youngest son Daniel, later to be the founder of the Massey Manufacturing Company, grew up. He had been just four years old when the family left New York state for Upper Canada. In the district in which the Masseys lived, there were no schools, so Daniel had to journey back to Watertown and live in his grandfather's home while he had three years of schooling. When he returned to Upper Canada, it was to help with the work on the farm and to look after his family when his father served in the militia. The War of 1812 set the British and Canadians against the Americans, and the elder Daniel felt it his duty to serve under the British.

With the end of the war two years later and the return of his father, Daniel began looking around for more adventure. The custom of the time required that he should work for his father until the age of twenty-one. But when he was nineteen, Daniel asked for his release. His father agreed on the condition that he should forfeit all claims to his estate.

Daniel rented land nearby. But instead of becoming a farmer, he chose to work for himself as a contractor and set about recruiting teams of laborers to clear land. The work was arduous but well paid. Newcomers were flocking into the fertile farm regions on the north shore of Lake Ontario, and the richer ones, whether they were immigrants from the United States, or from Britain, or elsewhere in Canada, were prepared

to pay high prices for cleared land.* The owners of the land
would pay Daniel and his work teams to cut down trees, and
burn and clear the underbrush; then they would settle the land
or sell it off. By 1830, Daniel and the men he employed
had cleared more than one thousand acres of land around
Grafton and the nearby town of Cobourg.

Daniel himself was a strong, hard-jawed man who combined
the rough appearance and habits of a laborer with a personality
that was in fact more temperate. Despite his scant formal
education, he had an enormous intellectual curiosity and a
practical and inventive nature. In clearing land, the timber
had to be laboriously trimmed and rolled by hand into piles for
burning. Daniel saw that it was simpler and more efficient to
use oxen, and was the first to do so. He was also a man of deep
and firmly held religious principles. He abhorred strong drink
and would not let his workers take whisky, even though on
one occasion this led them to walk out and refuse to help harvest
the crops on his farm.

In 1820, Daniel had married Lucina Bradley, whom he had
known since childhood. Like her husband, Lucina was descended
from American settlers who had come north; her grandfather
had sought sanctuary in Canada after their home in Illinois
had been burnt by raiding Indians. A year after their marriage,
a daughter was born. And on April 29, 1823, a son was born
and given the name of Hart Almerrin, in honor of his grand-
uncle in Watertown. In all, Daniel and Lucina were to have
seven children who survived infancy. Five of them were girls.
Of the two boys, the younger, William Albert, died at the age
of seventeen. Hart, the other son, was healthy and strong, and
the inheritor of his father's forceful personality.

Father and son were close, and Hart grew up working
beside his father. At the age of seven, he could handle a
team of horses and made weekly trips to the grist mill. Daniel,
meanwhile, devoted more of his time to farming and less to
the labor of clearing land. And he discovered a new interest
for his talents in the mechanization of farming.

The agricultural methods of the time distressed him. In a
letter to the newspaper in Cobourg in 1840, he wrote: "The
tools in use on farms all over the world are the same as those
used in the days of the Pharaohs. There has been a small metal
share added to the plough and some smart Scot has put a long
handle on the sickle a few years ago, and we have the scythe.
But the backbreaking chores of sowing, reaping, threshing and
cultivating, are still nightmares to most farmers."

---

* In this period, the population of Upper Canada (modern-day Ontario) was growing
swiftly. In the ten years after 1814 it rose from 95,000 to 150,000, and in the next
ten years it more than doubled.

Hart shared his father's interest in farm machinery. On a visit to Watertown, Daniel and his seventeen-year-old son saw some of the latest machinery including a mechanical threshing machine; they were so impressed they purchased and shipped one across the lake. The big barn at the Grafton farm became a workshop. The threshing machine was the first in Upper Canada. It required eleven men to work it, whereas threshing with flails required more than twenty.

In the countryside around Cobourg, and throughout the province, economic conditions were changing for the worse. Canada was still receiving large numbers of immigrants from Europe, but the arrivals who came ashore at the lake ports after the long ocean journey, or travelled overland from Montreal were often desperately poor. They had fled from crop failures and famines only to find conditions in Canada were scarcely better, and it was hard to find work.

The disasters in Europe followed the newcomers. In the 1840s, Britain had been forced by the famine and starvation in Ireland to repeal trade regulations that had, until then, favored Canadian farmers. Wheat and flour had been shipped to the mother country under preferential terms. Now the Corn Laws had been revoked by the British Parliament, ending the trade arrangements and striking at the livelihood of the farmers of Upper Canada.

The chief export of the colony had been flour. The dramatic falling demand for it had a chain effect among the communities dotted along the shore of the St. Lawrence and Lake Ontario; less flour to ship to the British market meant less need for flour mills and led to the impoverishment of the millers as well as farmers and agricultural workers. The depression in trade also hurt the foundry business. During the 1820s and 1830s, scores of small foundries had sprung up to supply farmers with ploughs and other implements, and the flour millers with the machinery for their mills.

Many small businesses were failing. Families were facing ruin and being forced to sell out to anyone who would offer them a good price.

For Daniel Massey, a pillar of the community and a comparatively rich man, the economic decline was a cause of great concern. But it also presented him with an opportunity. One of the businesses that had failed was a small foundry less than twenty miles from the Massey home, near the village of Newcastle. It consisted of two buildings, a foundry and a blacksmith shop. Its previous owner had supplied the neighboring farmers with a range of products, from ploughs to sugar kettles. Daniel was determined to buy the business. First, he sold his farm to Hart who had recently married Eliza Ann Phelps, of

Gloversville, New York. Then, aided by a partner, R.F. Vaughan, whom he later bought out, Daniel purchased the foundry and embarked on a new career. He was forty-nine years old, and the company he started was called the Massey Manufacturing Company.

The business prospered and Daniel soon had to think of expanding. He purchased a solid brick factory building close to the newly opened road from Toronto to Kingston. The two-storey building had a six-horsepower engine, a machine shop with one lathe, a cupola furnace of one-ton capacity, and was known as the Newcastle Foundry and Machine Manufactury. It had ten employees.

In the years that followed, it was no easier for Daniel to keep pace with the orders. He bought a bigger horsepower engine and added new machinery. And he took on more work-men. But Daniel was now in his fifties and, after a life of un-remitting labor, growing tired. He decided to ask Hart to come into business with him. Hart had been doing well and had settled down to the life of a prosperous farmer. He and Eliza now had two sons, Charles, born in 1848, and Chester, in 1850. When his father's request came, Hart rented his farm, moved his family into the white frame house beside the Newcastle works, and began to help his father. He was twenty-eight years old.

Hart's workday schedule at the factory showed him to be a Massey in the industrious tradition of his father and grandfather. He would start work early and continue through the day, overseeing the workmen and checking the quality of work in the factory, meeting customers, and travelling by horse and buggy to neighborhood exhibitions and demonstrations of farm machinery.

Active, concerned, involved in community affairs, deeply religious, all these were traits Hart acquired from his father.* But Hart's personality and character marked him as an altogether different, more worldly man than his father. Perhaps it was the advantage of a formal education, something his father lacked. Or perhaps it was Hart's judgment, his self-confidence, and his ability to organize others. Certainly he was a brilliant businessman, and he was consistently able to make the right decisions and to guide the family business as it grew larger.

Daniel came to recognize this. Hart wanted to import and sell the latest machinery, thereby beating their competitors, but his father was doubtful and preferred to take fewer risks and follow the lead of others. Eventually Hart's view prevailed.

* Hart's strongly held religious convictions led him into public life and service to the community; he was also a school trustee, the leader of a Bible class, a justice of the peace for Northumberland and Durham counties and, for a time, the district's chief magistrate.

He obtained the rights to two American patented machines and added to his catalogue Canadian versions of the popular mowers and reapers in use on U.S. farms; later these were replaced by a combined hand-rake reaper and mower. Hart had judged the market well. Moreover, he saw to it that at the Newcastle works the machines were well finished. The Massey family firm began to develop a reputation not only for innovation, but also for the excellence of its workmanship.

Within four years of inviting Hart to join him, Daniel had decided to retire from the business. One year later, in November, 1856, he died. He was fifty-eight years old. On his deathbed the doctors prescribed liquor as a stimulant and to help his breathing. Adamant to the end, Daniel said he would rather die than take strong drink. In the ten years since Daniel had founded the company, the business of the Massey Manufacturing Company had shown a fourfold growth and it was to continue to grow as rapidly for the next ten years.

A month before Daniel's death, an event had occurred in Cobourg which was in many ways symbolic of the new machine age, and the expansion it made possible. The Grand Trunk Railway, constructed to link the cities of Toronto and Montreal, reached Cobourg. Hitherto the business had been mainly a local one, and Daniel had been content that it should be. Hart, however, had grander ambitions. The coming of the railway was to hasten the fulfilment of these ambitions. From now on Massey products could be ordered and dispatched to distant points.

Change was also transforming the way of life of farmers as mechanization began to alter agriculture in North America and the world. The methods of the Pharaohs were giving way to the mower, the binder, and the tractor. These changes were occurring in the United States rather than in Canada. But Hart Massey, with his close connections and many relatives in the U.S., was aware of them, and kept abreast through tours of American plants and attendance at field trials.

It was from one tour that he returned with the rights to manufacture the Ketchum mower and the Burrell reaper. The mower allowed one man to cut large quantities of hay for feed for the horses. The reaper was the first machine to be used in harvesting. A few years later, the combined hand-rake reaper and mower was developed in the U.S. and brought to Canada by Hart Massey.

His own manufacturing plant was not the only one to borrow ideas from south of the border. Others were also doing it. In 1861, Canada had sixty-one companies in the farm implement business; by the end of the decade the number had grown to 271. One rival venture, which was to loom large in the development of Massey later on, was the Harris Company, founded by Alanson Harris, in Beamsville, Ontario. Started as a small factory, Harris was running his business in the same way as Hart. He had obtained the manufacturing rights for the highly successful American-made Kirby mower.

At the time, there was plenty of new business. The Massey and Harris companies were ahead of their rivals because of their willingness to take American designs, manufacture them, and create Canadian markets for them. The economic recession of the forties had given way to a period of prosperity. The American Civil War had driven up grain prices. And the infant Canadian farm industry was ignored by the bigger U.S. firms. American companies found it simpler to sell their designs to Massey and Harris than to attempt to wrest their markets away from them. Transportation between the two countries was still poor. Rather than try to sell to Canadians directly, the Americans stayed at home.

For Hart, the opportunities to expand his business were there to be grasped. And he was not a man to let opportunity pass by. Instead he rushed headlong into enlarging his factory and advertising for new business. He would provide farms with the latest machines and implements. And he would do so, he claimed, at the finest, most modern factory. It was in this thriving atmosphere, with plans being made to employ more skilled woodworkers, that disaster struck: A fire that broke out in March, 1864, gutted the factory and destroyed nearly everything. It was a tremendous setback but not an insurmountable one. Within a year, the factory was operating again while Hart had been busy selling implements even while the new factory was under construction and the new machinery installed.* When the foundry opened, Hart had orders for four hundred implements. It was more business than he could handle.

The first ever export of machinery overseas.

On Dominion Day, 1867, the first in the history of the new Canadian nation, the celebrations in Cobourg were as well organized as anywhere. At midnight, the church bells rang out. At noon, five companies of militia gathered on the common ground outside the town. Artillery were fired off. There was a torchlight parade in the evening and fireworks to end the day.

Within the Massey family, things were also coming to

* A newspaper advertisement placed by Hart at the time read: "The subscriber flatters himself that with his superior facilities he can compete with any establishment of a similar kind in Canada or the United States."

fruition. Hart's business and his reputation as a maker of farm implements had expanded greatly. So much so, that at the Provincial Exhibition held in Toronto that year, a committee of government officials had decided that the honor of representing Canada at the International Exposition in Paris should go to the Massey Manufacturing Company, of Newcastle. In Paris, the firm would be exhibiting its products with those from fifteen other countries. For Hart, the trip to France was to be a rewarding experience; it would open up new horizons, both for him and for his business.

The Massey family was made up of Charles, age nineteen; Chester, seventeen; Lillian, the only daughter, thirteen; and Walter, three. In the course of the year, Eliza gave birth to another son, Frederick (Fred) Victor, their last child. When Hart departed for Paris, he left the family and the firm in the hands of his eldest son, Charles. Educated at Victoria College in Cobourg and at the British-American College in Toronto, Charles had been accustomed to helping his father with the business and assuming responsibilities. A diligent worker in the tradition of his father and grandfather, Charles was, at the same time, a more social, gregarious personality. He had little of the reserve and stiffness of his father, and none of the austerity of his grandfather.

The Paris exposition was a great success for Hart. The Massey harvesting equipment won the first grand prize and two other gold medals. One of the medals was given for a field test in which Hart drove a hand-raking reaper. Designed by an American, Walter Wood, it was to European eyes, a great advance. It meant one man could do the work of two.* An admirer at the exposition was Emperor Napoleon III of France who declared that only Massey equipment would be used on his farms. Despite this pronouncement, Massey did not received an order from France.

One country that did respond to Hart's European trip was Germany. An order was received for twenty harvesting machines. Charles, who was a natural salesman, took charge of the shipment. The boxcars on the train carrying the machines were decorated with red, white, and blue streamers. The town band of Newcastle played German music. And the townspeople and workmen were on hand to provide a send-off for Massey's (and Canada's) first ever export of machinery overseas.

In some respects these triumphs were isolated ones. It would be many years before the company would build effective contacts, and the sales organization, needed to market its

* The combined mower and reaper could gather grain up in any terrain. It also cleared and deposited the grain in the field ready for binding.

products overseas. The small plant at Newcastle did not have
the capacity to handle orders from abroad. Still, Hart's journey
had shown that Massey products were of high quality. This,
combined with the innovative lead gained through the U.S.
patents the firm obtained, placed the company in the fore-
front of farm mechanization.

When he returned, Hart set to work with Charles to
increase sales in Canada. But their partnership was a brief one.
The constant exertions and heavy work had taken their toll
on Hart just as they had on his father before him. During
the summer, he took a long vacation with Eliza and the four
younger children, again leaving Charles in control. When he
returned to work, it was only to be advised by his doctors
that he should take a rest and reduce his activities. In 1870,
at the age of forty-seven, Hart decided to leave the business and
move to Cleveland, Ohio, where he had friends and relatives.*
As for the hard-won enterprise he had built up, Charles had
shown that he could manage things, and do it well.

A born promoter.

In any family dynasty, it is rare for the enterpreneurial drive
and spirt that animated the founder to pass on to a third generation. Charles Albert Massey was a very different personality
from his father and grandfather. He did not have Daniel's
shrewdness. Nor did he have Hart's ambition. Nonetheless, the
qualities he had were remarkably suited to making a success
of the family firm as it broadened its base and sought new
customers. Charles was a born promoter. Well educated,
he was more adaptable to new ways of doing business and
new ideas than either Daniel or Hart. His grasp of the tenets
of the machine age was surer.

Hart in Cleveland was not far off. There was a constant
exchange of letters and several visits. But Charles had taken
over, and was running things his way. When Hart departed,
he had changed the firm into a joint company, with capital of
$100,000. The Massey Manufacturing Company had Hart as
nominal president, with Charles as vice-president and superintendent.

Quickly, Charles began to make his imprint on the firm.
The economic conditions at the time were by no means favorable.

* Hart fully intended to retire in Cleveland; he had a fine home built for the family there
and became an American citizen. He also toured the southern states in the company
of a Methodist minister, giving generously to charitable institutions and churches.

28 Unemployment in Ontario was rising and prices were falling, with a large number of business failures. Under Charles' guidance, Massey not only survived but prospered. Sales rose by as much as 50 percent in the first five years of his management. And the work force at the factory was employed all year-round for the first time.

Charles managed to do this by coming up with new ideas and new products, and promoting them. His father's imports from the United States had often won their customers and achieved their sales by word of mouth, or in field trials. Charles would enter his machines in competitions. But he would also make sure that anyone and everyone connected to farming had heard about the latest Massey product. As an industrialist, he was perhaps the first in Canada to recognize the power of advertising.

Typically, a successful Massey product would be discovered in the U.S., and then launched in Canada. In 1874 a self-dumping wheel rake, known as a Sulky Horse rake, was shown to Charles, and its performance impressed him. It had been invented by an American named Sharp, and had an automatic dumping device that would save time and labor. Charles first bought the rights to manufacture and sell it in Canada. Then he marketed it using methods that would impress any modern merchandising manager.

"Sharp's Rake, no equal or no sale" read the advertising copy placed in the newspapers. A picture showed farmers laboring with old, cheap machines while a knowledgeable band of spectators looked on. "If you had bought a Massey, you would have had no trouble," declared one of them. In the foreground, an angelic lady carrying a floral tribute inscribed "Victory" was shown seated on a Sharp's rake. The Massey advertising campaign was one of the most expensive ever mounted. Another promotional gimmick was a letter written to the company by a ten-year-old girl. Sharp's rake was so simple to operate, wrote the precocious youngster, that she had been using it on the family farm since she was eight.

Charles believed that advertising was the best means of selling his products, and Massey advertising was widespread and inescapable in an era when most Canadians were farmers, and all of them were interested in the latest developments in farm mechanization. At one time the Massey catalogue came out with an announcement that there would be no end-of-season bargains because all Massey products were selling out so fast. Another time, when sales were slower, Charles decided to popularize his machines by giving them new, fancy names; farmers were offered the chance to buy the King of the Meadows, the Queen of Harvesters, and the Mighty Monarch of the Fields.

The success of the campaigns was justification enough for Charles' methods. The works in Newcastle had been expanded to a dozen buildings. Three thousand Sharp's rakes were being manufactured every year. And when Massey launched its first Canadian-designed harvester, the factory could not cope with the orders that flooded in. Massey was succeeding where others were failing.*

The great prosperity enjoyed by the company also meant that it was outgrowing the factory site in Newcastle. The land beside the Kingston Road first developed by Daniel could no more contain the rambling buildings. Faced with five hundred orders for its new harvester, the works could only cope with three hundred and fifty, even when the men worked day and night. For Charles, this was reason enough to move to larger premises.

When the decision to move to larger premises was taken, Hart came back from Cleveland. He was excited by the achievements of his son. When he had left, he had thought that Massey's best years were over and that trying economic conditions would make the survival of the firm difficult. Now, in the summer of 1878, much improved in health, he was returning to devise with Charles their plans for the future, and for a new factory.

There was no doubt in the mind of either of them where the new factory would be. When businessmen had achieved success in the small towns of Ontario and wanted to expand, they headed for the provincial capital of Toronto. A few years earlier, a store owner called Timothy Eaton had decided his ambitions lay beyond the town of St. Mary's. He had established himself in Toronto, in the process founding the Eaton department store chain. Now the Masseys were to move to Toronto.

Their arrival caused a stir. The great factory they were building was unlike anything in the province at the time. It stood on six acres on the western side of the city, bordering King Street. On one side were the railway tracks, with spur lines leading into the factory buildings, and on the other an ornamental garden and fountain facing the road and the streetcar line. The building itself was large and imposing, its three-storey redbrick facade contained the main factory and office building, with a foundry, blacksmith shops, and an engine and boiler house beside it. The Massey's new plant would house everything needed to manufacture farm machinery.†

Construction took place all through the summer of 1879.

---

* When called by the House of Commons to give an account of his business before a Select Committee inquiring into "the Causes of Depression in Trade and Commerce," Charles Massey stunned the MPs by telling them the company had had gross sales of $100,000 in 1875.

† Charles, with his taste for modernity, also decided to equip the factory with such daring new conveniences as steam-powered elevators, telephones, and sprinkler systems.

At the end of that year Charles and his wife Jessie, and their young family, left Newcastle and moved to Toronto. Meanwhile Hart, after poring over the plans for his monumental new plant, had returned to Cleveland for a time.

In the fall of 1882, at the invitation of Charles, Hart came back to Canada. For his return the family purchased a 25-room mansion on Jarvis Street. A solidly Victorian home in the most fashionable district of Toronto, it was to be the family home of the Masseys for many years.

Hart's return was encouraged by his son. The business was becoming bigger, and Charles realized he needed help with it. A year before, in a pattern that was to be repeated, Massey had acquired a struggling competitor, the Toronto Reaper and Mower Company, which had developed a new binder. This binder, which cut the grain, bound it with twine, and stacked it in the field, was one of the most revolutionary advances in farming. Harvesting time could be reduced, and a great deal of work saved. To sell the binder, Charles came up with the idea of selecting one hundred farmers to test it and publicize it. The farmers were enthusiastic. And the company, with a great deal of fanfare, announced plans to make 1,000 binders before the next harvest.

Rapid growth was also taking place in the company's markets. In Canada where the west was being opened up to settlers and farming, Charles arranged for a company to distribute all Massey's products from the city of Winnipeg. The results were spectacular. In less than ten years, the business done by Massey on the Canadian prairies went from $2,000 to $143,000.

All this growth and prosperity had made the Massey family very wealthy even as it had greatly enlarged their business so that it had become impossible for Charles to run it by himself. No thought was given to involving outsiders in the management of what had always been a family-run undertaking; instead, Hart and his younger sons, Chester, Walter, and Fred Victor, moved back to Canada. Charles' business achievements had made it necessary for the older, and younger, generations to involve themselves in an industrial enterprise rapidly becoming one of the largest in the country.

As it turned out, the presence of all the family in Toronto was needed because of tragedies, not triumphs.

Shortly after the return, Hart was struck down by a mysterious illness. Four doctors attended to him and pronounced that he would not live. But after a few days, he began to rally. Within a few months he had made a complete recovery and was working again at the factory. But his son was not to be so fortunate. In the winter of 1884, Charles was taken ill with

typhoid fever, and his sickness grew worse. Doctors tried to
save his life by giving a blood transfusion and then injecting
milk into his veins. It was to no avail. On February 12, Charles
Massey died at the age of thirty-five.

The Masseys under attack.

In its first forty years the Massey Manufacturing Company
had grown substantially, while remaining essentially a family
business with its roots in rural Ontario. There had been tremendous expansion. And the Toronto works, which employed
1,000 men, was testimony to this; but it was typical of the firm
that, when the move to Toronto had been made, the Masseys
had brought many of their Newcastle workers with them.
Some fifty families had made the move, loading their belongings
on wagons and travelling to the city to find new homes.

Under Charles' direction, the spirit of paternalism, which
guided the new industrial enterprises of the era, was given
full rein. Massey informed its workers of the great events of
the day through the *Massey Pictorial, Massey's Illustrated*, and
a magazine called *The Trip Hammer*. There was a library
association, lecture halls, and reading rooms for workers in
the plant, and Charles' own love of music led to the founding
of the glee club and the Massey Cornet Band. All this was
normal in the new industrial era. Owners and workers thrown
together in a large industrial organization, and living for the
first time in a big city, sought to recreate for themselves
the homely social life of the small towns they had left behind.*

Charles' death put all the responsibility for running the firm
on Hart. The second son, Chester, was appointed a vice-
president of the firm. But while Chester had inherited the
intellect and the Methodist faith of his father, he had little
taste for the business world. Because of this, Hart reached down
to his third son, Walter, who agreed to cut short his university
career in Boston and return to Canada. Walter was cheerful,
and talented in his studies. He had some of Charles' charm and
verve. And he had the curiosity, and the delight in new things,
of his grandfather Daniel. However, he was youthful and
inexperienced. And so, as the company moved into the second
half of the 1880s, it was Hart, now in his sixties, who remained
its dominant personality.

The challenges that lay ahead were basically two. First,

* The population of the city of Toronto had risen from 30,000 in the 1850s to 180,000
  by 1890.

the company had to continue to grow and widen its markets. In this respect, Hart, as a knowledgeable and clever business-man, could be expected to succeed. The second challenge, was a more complicated one for Hart to meet, it was to change with the times and respond to the needs of a new age. The union movement was growing. Demands were being made for higher wages and financial aid for disadvantaged workers. Old standards of morality were being relaxed. New attitudes towards labor were being expressed which, more and more, were finding their way into the newspapers and other forums of public opinion. To all this, Hart Massey responded like a man who was both blind and deaf. He neither knew nor understood what was happening, but he was resolutely opposed to it.

The start of the trouble came when the company began paying its workers on a piecework basis during a period of flagging economic fortunes. Though the Massey workers were grateful for holding their jobs when other companies were cutting employment, they were convinced the system was unfair and the company was cheating them.

When their grievances were not listened to, some of the workers went to a union, the Assembly of the Knights of Labor, which presented two demands to Hart; wage rates for each worker must be posted in the factory, and wages must be paid every two weeks. Hart not only refused to discuss the matter, he fired all the men who had joined the union's committee and voiced the complaint. He was asked to reinstate them, but he would not. Two hundred workers then went on strike. And in the days that followed they were joined by two hundred more.

The labor unrest came at a time when Hart and his family were already under criticism. In the economic depression, the richer citizens of Toronto had become a target for the poor, the dispossessed, and those who were simply envious. And close to the pinnacle of wealth were the Masseys. They were accused of living luxuriously, of exploiting their workers, and of flaunting their wealth when they made gifts to the city. The farmers, too, were in full cry. They considered prices of farm implements excessively high. One indignant writer to a newspaper pro-claimed: "From Masseyism, Good Lord Defend Us."

Considering the scrupulousness of Hart in his business dealings, his integrity, and his hard work, much of the criticism was unjust. Still, it continued to mount, and the public mood — whipped up by the strikers — began to turn ugly. When they when to the plant, Hart and his sons had to be protected by police.

To Hart, the unrest and the threat of violence came as a shock. But it only served to reinforce his determination. He would not give in. There were obligations which bound an employee to an employer and, in his view, the strikers had violated them. Young Walter tried to prevail on him to compromise. Instead Hart ordered that those who were refusing to work should be paid off, and he advertised for new workers. The strategy helped to give him victory. There were so many unemployed willing to take jobs that the workers had to return or lose their livelihood.

Nonetheless, Hart had created a considerable undercurrent of bitterness and hostility. Among the public at large, it continued to ferment for many years. The Masseys were branded as uncaring and unsympathetic capitalists and oppressors. Hart in his autocratic way could not have been less concerned. In the aftermath of the strike, he tried to press home his advantage by forming an employers' group to fight the Knights of Labor. Hart wanted to win an agreement from the city's other big employers that no striking Massey employee would get a job with them. He also donated money to the Police Benefit Fund as a gesture of appreciation for the protection his family had been given during the strike. The gift was returned by the police on the grounds that it was unethical.

These clumsy moves were the actions of a man under stress. Hart had little comprehension of and no sympathy for the social forces that were at work altering an earlier paternalistic style of industrial management. The Knights of Labor had been beaten, and their strike had been a failure. But they had not gone down to permanent defeat. A few years later, they were back and agitating for change. The staff in the office campaigned for a shorter work week. Afraid of Hart, they presented their petition to the mild-tempered Chester instead. The rumblings that Hart had set off by this opposition to strikers, and to anyone else who earned his moral condemnation, continued to rankle. The Masseys and their company were powerful. By 1893 they were the manufacturers of half the farm implements sold in Canada. But they were no longer loved.

Hart builds an empire.

The power and authority that accrued to Hart Massey and his company came from the continuing expansion of his business. And this was a cause for much of the hostility that his name evoked. The early 1890s were a time of economic depression, but to Hart, whose business was large enough to be unaffected, it was a time of opportunity. When weaker companies faltered or got into financial trouble, he was in a position to purchase them or merge with them, and he could buy in at a cheaper price. In the early years of the decade, Hart preyed on his competitors, looking for their weakness and striking deals to his own advantage. Hart's single aim was to enlarge his own empire, and he was quite ruthless about it.

The biggest merger of all was with A. Harris, Son and Company Limited, of Brantford, the firm that from its original home in Beamsville had developed successful products and had challenged Hart in his early years by picking up the rights to as many American patents as he did. The career of Alanson Harris and his son John had paralleled the Masseys; the father leaving his farm to start a sawmill and foundry, later being joined by his son who persuaded him to begin manufacturing farm implements. And they had been successful. The great Massey sales festivals and promotions, the delivery day parades, and the glossy catalogues, devised by Charles, were put together in an effort to beat the Harrises. The Masseys had drawn ahead. But the Harris Company had just announced a breakthrough that was winning them new markets in Canada and abroad.

This was the open binder, a machine that could cut and bind grain of any length. It meant that for the first time a farmer could buy a machine which could be adjusted to harvest in any crop conditions. Hart realized that he had been beaten by a more advanced product. Rather than face the prospect of losing sales, he called for a meeting with Alanson Harris and proposed that the two companies, the largest and second largest farm machinery companies in the country, should merge. On May 6, 1891, the Massey-Harris Company was established. Under the terms of the merger, all the patents, production methods, and facilities of the two companies were put under Hart's direction. Wily and aggressive, he had emerged as the chief of the biggest industrial empire in the land.

After this victory, Hart acted like a good general and turned to pursue his weaker opponents. The merger had alarmed another rival, the Wisner Company, which had enjoyed a close business association with the Harrises in Western Canada. Together they had competed against Hart Massey. The Wisner Company tried to safeguard its interests by joining forces with another manufacturer, the Patterson Company of Woodstock, Ontario. The threat that this posed to Hart Massey was short-lived. By December, 1891, he had won over Wisner and the Patterson brothers to the idea of becoming junior partners in the Massey enterprise.

The acquisition widened the range of Massey-Harris products. Both the bigger companies had specialized in harvesting machinery. The smaller companies had offered a full range of farm implements. The process of consolidation went a step further in May of the following year. Hart was able to announce

Company, further widening its product line. With the sale went
the exclusive right to market all Verity products.

Each of these companies was led by a successful pioneer.
In taking them over, Hart was careful to take account of the
susceptibilities and wishes of the founders. They were permitted
to continue managing their companies, or given sales rights to
certain territories, or invited to join Hart on the board of
directors. Still, it was a market takeover he had masterminded,
and it was hugely successful. In the first years after the companies
had been combined, their sales rose by 20 percent. William
Verity and Wareham Wisner died a few months after the merger,
and Alanson Harris died two years later. As a result Hart
was at the pinnacle of a massive industrial enterprise, producing
farm machinery in Ontario and selling it in Canada and abroad.*

Growth had not come just from the expansion of farming
and farm mechanization, or even from the program of aggrandizement which had left Hart in charge of a whole industry. It had
also come from a remarkable effort at developing sales
overseas. The first prize-winning foray had been to the International Exposition in Paris in 1867. Others followed, including
the successes that Massey machines achieved at the International
Exhibition in Antwerp in 1885, and the Indian and Colonial
Exhibition in London one year later. From these trade fairs
and exhibitions, the company made its first halting steps into
the international arena, establishing sales agents, and beginning
to export its products worldwide. The process was started off
by the Massey and Harris companies separately, and it received
a considerable push from their amalgamation.

Though there was little in his background to suggest he was
an internationalist, Hart saw the benefits to be derived from
sales abroad. Facing fierce competition at home, he had during
the 1870s and 1880s realized that his harvesters were superior to
those manufactured in Europe. It followed that, if he could open
up sales in Europe, then profits would be substantially better
than they were in Canada. The first connection was with
Britain, and it was here and then on the Continent, that Hart
concentrated his efforts. He did not attempt to sell into the
U.S. Many of his designs and products had originated there.
Moreover, the U.S. companies would have given him too much
competition.

The steps that Hart took to explore overseas markets and
develop sales were impressive, and they marked him as an

* When the annual Massey-Harris picnic was staged in Newcastle a few years after the
mergers, it required six trains to take the employees and their families to the site. There
were 3,500 people present, three times the number that had attended the Massey
company's fortieth anniversary outing in 1887.

outstanding businessman. Hart's initiatives showed the way for Massey-Harris and the same policies were carried on by his successors. Under Hart, Massey-Harris was on the way to becoming a multinational company, sustained as much by the profits it derived overseas as from its base in Canada.

For Hart, however, there were tasks not connected with business that he wanted to accomplish. Tragedy had struck again a few years earlier with the death of his youngest son, Fred Victor, at the age of twenty-three. In some respects, Fred Victor in his youth resembled Hart in his middle age; he was an intensely religious young man and had a desire to contribute to the public good. He had spent much of his time working among the poor in Toronto's missions. For Hart, the death of a second son was a devastating loss. And it reinforced in him the desire to spend his remaining years in the service of the church and in contributing to the welfare of his fellow man.

Having decided to devote his energies to philanthropy and his time to public works, Hart Massey would in a brief space of time make a contribution that has seldom been equalled. As soon as the new Massey-Harris Company became established, Hart turned his attention to charitable and educational activities. At one point Walter argued with his father about the company's failure to move into the U.S. and improve its sales effort (the criticism was to be one that would recur throughout Massey's history). Hart ignored his son. He had other more pressing matters at hand.

Part of his urgency was the realization that he did not have long to live. Another reason to move swiftly was personal; he had felt challenged by the death of a son who had been devoted to his own Methodist principles and who might — if he had lived — have contributed to the betterment of society. Hart was determined to create a legacy for Fred Victor. And he would also commemorate his eldest son, Charles Albert.

In memory of Charles and his love of music, Hart conceived of the building of a concert Hall for the people of Toronto. The cornerstone of Massey Hall was laid by his grandson, Vincent, in September 1893. Nine months later an ailing Hart Massey was able to present the keys of the building to the mayor.*

To honor the memory of his youngest son construction began on the Fred Victor Mission. The building, with accommodation for over 200 men, was completed in 1894 and given to the city.* In addition to these donations, Hart supported religious and educational groups and hospitals.

As the final testament to his purpose, he was to leave a will that was a truly remarkable document. From an estate totalling $2 million, the amount willed to his family amounted to $250,000. Of the remainder, $750,000 went to specific religious, educational, and charitable organizations, while $1 million was to be disbursed under the direction of his executors and was to provide the funds for a family foundation, the Massey Trust.

In 1894 Hart had become ill. He was never to fully recover. Nearly two years later, on February 20, 1896, Hart Massey died at the age of seventy-three. Not far from the Jarvis Street home in which he lay dying, a performance of *Haydn's Creation* was being staged at Massey Hall. When the oratorio was completed, the Toronto *Globe* reported that "the conductor, in brief but fitting terms, announced that the builder of the hall was no more, and the entire assembly rose as the great organ pealed forth in the most effective manner Handel's Dead March from *Saul.*"

3

Going
International

*The company which Hart had built up had
already outgrown Canada, and could not let
itself be confined to purely national considerations.*

Frederick Isaiah Massey was from the American branch of
the family, a cousin of Hart Massey. In his younger days, he
had been a captain in the Union Army and had commanded a
company at the Battle of Gettysburg where he was wounded in
action. After the war, he had become a businessman and treasurer
of the Dubuque Iron Works until a chance meeting with Hart
had led to his appointment as European manager for the Massey
Manufacturing Company, based in London, England.

It was a position that Frederick Isaiah was to hold for many
years, but in the summer of 1889 he had been in Europe only
a short time and had found the job a frustrating one. Hart had
a great belief in trusting the most important tasks only to other
members of the family or close friends. He had already perceived
the advantage to be reaped from cultivating sales in Europe; not
only were prices high and profits good, but Europeans were
agog with the wonders of the machine age and prepared to
reward the makers of the latest miraculous invention with prizes
and honors. Hart's machines, as he explained to Frederick
Isaiah, were not exactly new inventions but they did incorporate
the best design ideas of the most modern American equipment.
Hart was convinced that Massey farm machinery could sell
around the world.*

These were the arguments advanced to persuade Frederick

---

* An early success for Hart had been at an International Exhibition in Antwerp where
the city's agricultural commissioner, Baron de Gruben, took a Massey harvester and
tested it out on a field of oats. He awarded the machine a Gold Medal and declared it
superior to all other makes.

Isaiah to take up the post in Europe. First Hart had won over
the board to the idea of a London sales office, then he had
talked in his deliberate passionless way to his cousin. Frederick
Isaiah — he was known to members of the family simply as
"Fred Eye" — left for England in 1887. With him went instructions
to contact a Scot who lived in Paris, James Stuart Duncan. Hart
had met Duncan in London the year before and had been
impressed.

James Duncan was a young man in his mid-thirties who
had spent several years setting up agencies to sell farm machinery
manufactured by the D.M. Osborne Company of Auburn, New
York. He had travelled as far east as the steppes of Russia and
had managed to win Osborne a thriving business in the Mediter-
ranean countries, in North Africa, and in the Balkans. Travel in
those days was by train. And travelling conditions were often
insanitary and bad for the health.

When Duncan encountered Hart Massey in London, he
had been advised by his doctors that he should settle down and
travel less. He told Hart that it was possible for him to open
his own business in Paris and import and distribute farm
implements from there and he would be willing to do it with
Massey machinery. No deal had been struck. But when Frederick
Isaiah journeyed to Europe, it was with the idea of taking
Duncan as a partner in the venture he was about to start up.

Other contacts had been made by Hart. At the Indian and
Colonial Exhibition, he had met importers from Australia,
Argentina, Chile, and Russia. Armed with the conviction that
any job must be done thoroughly, Hart had recruited another
old friend, Charles MacLeod, to open an office in Australia.
He was also preparing to send his sons Walter and Fred Victor
on a world tour that would combine business with pleasure.
Walter had shown signs of fatigue and sickness. His father
considered the trip would do him good. It would also help
him in his choice of places to open sales offices, and gain new
business.

While other continents held out some promise, the biggest
and greatest hope had always been Europe. It was here that
Hart Massey had won awards. Both he and Charles had used
those awards skilfully. They had boasted about their triumphs,
including them in catalogues and brochures, and employed
them as advertising weapons against their adversaries.

Once in Europe, Frederick Isaiah discovered the situation
was rather different. In spite of the medals and awards, Europeans
were a long way from recognizing the name of Massey, or the
distinctive white and gold colors emblazoned on its machinery.
There was no doubt that it was possible to gain a market in
Europe but it looked, at least to Frederick Isaiah, as though the

process would be exceedingly slow. The British were at the time the biggest exporters of farm machinery to the rest of Europe. When Frederick Isaiah visited Duncan in Paris it was to find him acting as agent for two British manufacturers. Duncan assured him he was anxious to distribute American and Canadian products, too.

For Frederick Isaiah the challenge was to raise the esteem in which Massey harvesters were held, and get the company's name better known. In the summer of 1889, he was sure he had discovered a way of doing this. He wrote urgent letters to Hart, begging him to enter a competition which had just been announced by the French ministry of agriculture. A four-day contest for binders, the World Self-Binding Field Trials were to be held as part of the Paris Universal Exposition of that year. Hart was at first reluctant, but he eventually agreed that a single light binder would be shipped to France to compete in the trials. With it travelled William F. Johnston, superintendent of the Toronto works, who would represent Massey in the field.

The exposition, for which the Eiffel Tower was constructed, was a famous success. Twenty-five million visitors flocked to attend. All through the summer, the crowds thronged the 250-acre site in the centre of Paris where pavilions displayed the newest wonders in art, science, and industry.

On view, along with other farm machines, were the Massey products that had been shipped to Europe. For Frederick Isaiah the important event of that summer was not the activities on the Champ de Mars and the Esplanade des Invalides but the binding trials that were to take place on a rich landowner's farm at Noisiel, eighteen miles outside the city.

Fifteen contestants, representing the leading farm machinery makers in Europe and North America, lined up on the first day of the trials. Beforehand they had drawn lots for the most favored position. As they moved through the field, each of them was to be followed by a judge who would take note of the evenness of the cutting and binding, the time taken, and the number of men and horses needed for the task.

The Massey entry, with Frederick Isaiah and Hart looking on, started off alone. There was no spare operator to back up Johnston, and no mechanic. The Toronto binder required only two horses to draw it. All the others needed three or more. After thirty seconds, Johnston stopped his machine, and it looked momentarily as though the Massey entrant was out. After making a minor adjustment, Johnston remounted. He did not stop again.

In the first half-hour of the opening day, four of the fifteen contestants dropped out. Johnston finished in one hour and six minutes, and he cut his section without missing a single sheaf.

and many of the other contestants were still cutting and binding three hours later. On the second day, the story was repeated. If anything, the superiority of the Massey machine was more apparent. Johnston completed cutting his section of oats in fifty-five minutes, without stopping. His nearest rival needed a change of horses, and took two and one-half hours.

The elation of the Masseys could not have been greater. And it was shared by William Johnston. After his second triumph, he drove past the cheering spectators, dismounted to shake the hands of those he knew and, in a gesture of bravado, headed his team for an unharvested section considered too difficult for a machine to cut, which had been left over from the previous day. One hour later he had cleaned it up entirely.

On the third day, President Carnot of France visited Noisiel, and the Toronto binder was chosen to bind the sheaf that was to be presented to him. In a final competition, the Massey machine won another prize as the only binder light enough to be drawn by two horses. For all this, the company was awarded an elegant bronze statue classified as a "grand objet d'art."*

In the years that followed, the decision to enter the Noisiel field trials was shown to have been the right one. Sales to England and France rose dramatically. So, on the basis of testimonials from Europe, did sales to Australia. Massey's international reputation had been enhanced and it was to be given momentum following the merger with the Harris Company.

From this period onwards it was a common sight to see trainloads of machinery being shipped by the Canadian Pacific or Grand Trunk railways, westward or eastward from Toronto, en route to foreign markets. The firm had always liked to boast about its successes and awards. Hart and Walter were to make much of the new cosmopolitanism of their business. The four Toronto newspapers would be invited to see twenty-one freight cars leaving on the first leg of the trip to Australia. On each car was stamped the factory's emblem and the words "I am full of Massey-Toronto Harvesting Machines. Rush me along, the farmers are waiting. Loaded for Australia."

The Toronto factory was a large one, and the firm — when it combined with the Harris Company — had total sales of $3.3 million by 1893. This was a colossal turnover, and one that could only be supported if overseas markets were developed further.

---

* The triumph was marred for Hart. He had left France before the awards were given on the fourth day having been told that his son Fred Victor was dangerously ill in Toronto.

In his quest for international recognition, Hart Massey set himself apart from his contemporaries. Moreover, his quest put him at odds with the government of Canada in the years following Confederation.

If there was a single issue that divided the country, one on which newspapers editorialized endlessly, and politicians fought, won or lost elections over, it was the issue of the tariff. The nation had been born and brought into the world bawling and fussing about tariffs, and the degree of protection the national policy of Sir John A. Macdonald would provide for local industry. For Canada to survive, its industries must be succored by high tariffs — so the argument went. Only then could a nation strung out along a rail line that ran for thousands of miles from east to west resist the gravitational pull of trade with the U.S. An artificially created country needed artificial barriers to keep itself intact.

The tariff was applied to all industry, and it was applied to all imports whether they were raw materials or manufactured goods. Protection was universal. It was given to those who needed to be protected, or felt they did, and those who bitterly opposed it.

Among the opponents none was stronger than the Massey family. To them, the tariff was a hindrance. Most Canadians might favor high tariffs, but for the Masseys their own self-interest made them stand out against them. The farm machinery industry was not bothered by competition from the U.S. The American companies were content to sell their designs rather than market equipment, and leave the market to the Masseys and Harrises. They were also wary of entering Canada because of the advantage Canadian manufacturers had of being able to buy good quality steel cheaper from Britain. The same machines and implements could be made and sold less expensively in Canada.

The tariff did not protect the farm machinery industry; on the contrary it added to the cost because it was levied on raw materials that came in from the U.S. In the 1870s, when Confederation was only a few years old and Massey still a small Newcastle-based family business, Charles was pleading with the House of Commons in Ottawa not to raise the tariff above 17.5 percent. Other industries, however, needed protection. And in the economic downturn of the late 1870s and early 1880s, the clamor mounted. Tariffs jumped first to 25 percent and then to 35 percent.

To Hart Massey, this policy was short-sighted. If flew in

the face of his new-found internationalism and cut into the profits he could possibly expect from foreign sales. He wrote to Mackenzie Bowell, the minister of customs; he presented briefs to John Carling, the minister of agriculture; he importuned Sir John A. Macdonald himself. Hart's argument was that if costs were to be raised in Canada because of tariffs, how could Canadian farm machinery makers go out and win export contracts? Would not their business in Europe and Australia — which they had secured in the belief that it was for the good of the nation — fall prey to the Americans and others?

Hart's persuasion eventually won him a partial victory. At a crucial time, following the merger with Harris, the tariff was reduced to 20 percent. It was to remain at this level until 1907, and thereafter be lowered, and it in no way became an obstacle to the expansion of Massey abroad which continued all through the period. Still, the issue of the tariff would not go away. It would appear and re-appear in different forms; sometimes in the guise of free trade with the U.S. (as in 1910), sometimes in the form of trade wars and restrictive policies (as in the 1930s). Always, it would strike at the heart of the Massey-Harris operations. The company Hart had built up had already out-grown Canada, and could not let itself be confined to purely national considerations. Its interests were served by free trade around the world.

The controversy over tariff policies also taught the Massey family that it counted to have influence with politicians. The Masseys did not lack for political friends, nor did Hart put much restraint on his use of them. After the victory of the Toronto binder in Noisiel, he proposed to the minister of agriculture that the government should give publicity to Massey's and Canada's triumph. He was politely told that, if he wanted the public to know the good news, he should take advertisements in newspapers. On a more serious level, the Masseys and some of their industrialist friends exerted enough influence to keep tariffs lower than they might have been otherwise, and lower than the public often demanded. Whether this made Hart — and, after his death, Walter — champions of free trade, or self-serving entrepreneurs, is questionable. Certainly in Canada in the 1880s and 1890s, none could deny the primacy of Hart Massey as an industrialist and a national figure. When he died at his home on Jarvis Street, the family funeral service was attended by Mackenzie Bowell, then prime minister of Canada.

The influence the Masseys managed to exert on the tariff had a benign effect on their enterprise. The company now relied on overseas business, and each year increased this reliance, as sales were booked by agents around the world. There were Massey machines at work throughout Europe, including czarist

Russia; in Turkey and the Middle East; in South Africa; and in South America and Australia.

The greatest successes came in the territories covered by Frederick Isaiah and James Duncan. England and France were both major markets, particularly France. At the turn of the century, a farm depression seriously affected the industry in France, so seriously that it undermined Duncan's distribution business and pushed him towards bankruptcy. Massey was by then his most important supplier and insisted on being paid. When Duncan could not meet his commitments, the Masseys took over his business and made him stop selling competing products. Then they rehired him as the manager. These tactics might have soured Duncan to his new employer but, as an upright Scot, he was chagrined by his financial difficulties and willing to go along with the new arrangement and honor it. In fact, Duncan did as conscientious a job for Massey as he had done in running his own business, and the French market rapidly became the company's biggest export outlet.

In Europe, mechanization was growing not only in farms close to the towns but also out in the provinces. The pattern was the same as it had been in Canada, though with a greater degree of dislocation. In some areas, the arrival of the Canadian binders would be greeted with hostility; they were a threat to the traditional way of life in the countryside, the scything and tying of the wheat by hand. Still, progress was at hand in the small towns and villages. Typically, on a market day in some distant French locality, James Duncan wearing his grey top hat and morning suit could be found accepting a gold or bronze medal after his Massey-Harris machine had been judged the finest in the day's field trials.

At the close of one century and the opening of another, progress did indeed seem to count for a lot. Trade was expanding; tariffs were being lowered; and there was a general air of prosperity.

In Toronto, Massey-Harris was going through a transition period. Its founder had died, to be succeeded for five years by his son, Walter. Then in 1901 the bright and purposeful Walter had contracted typhoid fever and died. He was succeeded by his only surviving brother, Chester. In the process of these successions, much of the drive and dynamism that had been instilled in the company was lost. The forcefulness of Hart, and his two business-minded sons Charles and Walter, was missed. Nonetheless, the path they had set for the company was not deviated from. In an era of prosperity, Massey continued to expand at home and abroad.

But problems were building. And they were certain to loom very large for a company which had ambitiously moved into

world markets, linking its profitability to sales in foreign lands.
Markets in these regions would not be buoyant forever, and ahead lay a troubling historical period; the Great World War, the farm recessions of the 1920s, and the Great Depression. But at the dawn of the new century Massey-Harris continued to advance, unaware of the turmoils to befall it. Its global business seemed solid, the result of many years of painstaking work. All those bronze and gold medals were symbols of a success that no Canadian company had ever enjoyed before.

4
The Family Sells Out

For the firm that he led, however, Vincent Massey's sally into politics was to prove a turning point. Initially it was merely a source of conflict and confusion. But what was to grow from this was an ultimate separation between the Massey family and the company, the end of a family dynasty and of an era.

Typhoid fever, the disease that struck down Walter Massey, had also killed his brother Charles seventeen years earlier. When he died at thirty-seven, Walter was only two years older than Charles at the time of his death. Since Hart's death, Walter had been the driving force in the firm. With the death of Walter, Chester Massey was Hart's only surviving son, and while he had worked in the office, he had never been associated with the rigors of business. Chester had suffered from poor health since his youth and was unable to read without getting severe headaches. In Toronto's variable weather, he had the habit of going out in three overcoats, taking one off after another as he got warmer.*

In the aftermath of Walter's death, the directors of Massey-Harris, including Chester, met in the King Street offices to choose a new president. The discussion lasted for several hours, during which time harsh words were spoken and tempers became frayed.

The directors were no longer family members. The expansion that had taken place ten years earlier had brought onto the board men from the Harris Company, Verity Plough, and the Bain Wagon company among others and, although they had not opposed the Masseys during the lifetime of Hart and Walter, they had no particular allegiance to the family. Nor did they

---

* Despite constant worries about his health, Chester was to outlive all his brothers, his sister, and his wife Anna Vincent; he was seventy-six years old when he died in 1926.

consider Chester sufficiently involved in the business or know-ledgeable enough to occupy the president's chair. Far better to choose one of their own, a farm machinery man who had been brought up close to the soil in southern Ontario. With all this Chester was in agreement. And so was his wife, Anna Vincent, who was frightened about the affect the presidency would have on his health.

As the argument raged on, it became clearer that the disputants, those who rejected Chester, could not find any other candidate on whom they could agree. While none of them had a high opinion of Chester's business abilities, all liked him personally. So it was that at the end of the afternoon, another Massey emerged as president. The choice of Chester was a compromise. The directors simply could not bring themselves to settle on a stronger personality for the time being. For his part Chester accepted out of sense of duty and with the idea that his time as president would be as brief as possible.

Less than two years later, Chester stood down as president and was given the title of honorary president. His wife Anna hailed the event, describing the summer of 1903 as the period of Chester's "emancipation." With his business concerns behind him, Chester set off with his family, including his two sons, Vincent and Raymond, for a tour of Europe.

The first outsider in a line of strong personalities to run the company.

The departure of Chester did not signify the end of the family's involvement in Massey-Harris; that was to come later as a result of young Vincent Massey's political ambitions. Nonetheless, the directors who met to appoint a new president were looking for the first time for an outsider, a nonfamily member, and this surprising development was occurring less than seven years after the death of Hart Massey.

Their choice was the firm's general manager, Lyman Melvin-Jones. Autocratic and overbearing, he was to be the first in a line of strong personalities to run the company. In a sense, he was everything Chester was not; an experienced farm machinery man who had started his career with Harris in Brantford and progressed rapidly, rired by driving ambition. He had gone to Winnipeg to open a western office for the Harris Company and been so successful that, within a few years, he was elected mayor of the city. From the west, he had peppered the head office with suggestions about how machines in use on the Prairies could be improved. So many proposals had come in that Harris recalled him to Brantford and put him in charge of developing new products. From this base, Melvin-Jones had launched the phenomenally successful open-end binder. And when this prompted the principal of the firm's rival, Hart Massey, to talk about a merger, he had emerged as the first general manager of Massey-Harris.

Melvin-Jones' leadership style required that others obey his instructions to the letter. As he was knowledgeable and experienced in his field his decisions were seldom questioned, nor was he himself called to account for them. In his iron will, the other directors may have seen something of the personality of Hart. But what they had not considered, and what Melvin-Jones' period in charge of the company was to render crucial, was the need to groom others as possible successors and to develop a strong team of executives. The president wanted things done his way.

In the years leading up to the First World War, North America was moving west and the farm machinery industry was part of the great migration. Ontario in 1901 had grown three-fifths of all Canadian grain, but in the years that followed, many Ontario and immigrant farmers went to the west. The farm industry moved with them. Agents and distributors would stage delivery day parades in small prairie communities, throwing in a free dinner at the local hotel for farmers on the day they received their new Massey-Harris machines. While the company was unquestionably doing well in the west and ranked as the largest in the country, it was not without competition. As Canada grew and the west opened up, U.S. manufacturers began to take an interest. No more was Canada a market that did not merit attention.

The Americans first had to get around the tariff. This they did by establishing factories in Canada. In 1902 a merger of five companies had created a new Goliath, the International Harvester Company, which controlled nearly all the U.S. market for binders and four-fifths of the market for mowers. In the following year, the company started building a plant in Hamilton, Ontario, to compete with Massey head-on. For executives at the King Street head office, this was a serious matter. Their company was big, but although it had substantial sales in other markets, it had not penetrated the U.S. And compared with International Harvester, Massey was many times smaller.

Worse was to come. The enthusiasm for tariffs had in the space of a single generation reversed itself and a totally different political creed was now being put forward. Reciprocity in dealings between Canada and the U.S. or free trade, became the prevailing dictum. Massey had sounded the alarm when high tariffs had made the prices of its products uncompetitive in foreign markets. Now it faced the prospect of its Canadian market, which accounted for half of the company's sales, being swamped by the Americans. Since U.S. firms were much larger, they could manufacture and sell their machinery at a lower price. Not surprisingly, the idea of reciprocity was opposed by Melvin-Jones. But, since there seemed a high probability that it would come about anyway, he decided to act first and worry about the consequences later.

U.S. firms were able to offer lower prices because of their larger market and the economies of scale they enjoyed. Hitherto Massey had been inhibited in trying to compete in the U.S. because of steep tariffs going the other way. The solution, decided Melvin-Jones and the other directors who remembered the Canadian mergers of the 1890s, was simple and direct. Massey-Harris would buy an American firm. With this in mind, the directors travelled across the border to look at the premises of the Johnston Harvester Company in Batavia, New York. Johnston had only just missed being amalgamated into International Harvester. Now it became part of Massey's first foreign venture into manufacturing.

The decision was a disastrous one. First, reciprocity did not come and so the move had been unnecessary. Second, Melvin-Jones had completely misjudged his acquisition. Johnston's strengths were also Massey's strengths; it was a company that had built up its export business, and was not strong in the U.S. market. As a result, the purchase of Johnston brought few cost savings. Wages were slightly higher in the U.S. so farm machinery made in the Johnston works could not be sold competitively in Canada. Nor, with its weak marketing organization and small size, could Johnston increase its sales in the U.S. For a few years, the U.S. company was able to push up exports and the error of the acquisition was not apparent, but by the 1920s staggering losses were being reported.

At the same time, there were tough problems that Melvin-Jones could do little about. The war in Europe, and the invasion of France by the Germans, had ruined a large portion of the firm's export business. Nevertheless, the things that were within his ability to influence, such as the move into the U.S. and the challenge of maintaining a modern line of products, were mishandled badly.

Massey was losing ground to American manufacturers at home. Its effort to mount a counter-attack in the U.S. had proved to be a feeble one. It was also suffering a setback from the appearance and huge sales of the first farm tractors. The market was growing at a tremendous rate between 1910 and 1920, as farmers in the U.S. and Canada were won over to the concept of motorized farming.* During all these years, the company that Hart Massey had built on the basis of product leadership and technical innovation failed to respond or compete in the market in an effective way.

First, the decision was made to import a tractor from the Bull Tractor Company in the U.S. The lumbering three-thousand-pound machine was of poor quality and unreliable and had to be withdrawn frm the market. Then, the company used its plant in Weston, Ontario, to make a tractor designed by a Chicago firm. But this too failed. After producing a number of models under the Massey-Harris name, it was decided to close down the operation. Competition was fierce since all tariffs had been removed from the most popular lines of imported tractors. The company could not match it rivals and had been forced to give up the attempt.

All these difficulties could not be laid at the feet of Sir Lyman Melvin-Jones. But to a large extent he had, through his egocentric ways, created a climate of misfortune. It was hard to see whether the setbacks were caused by his own wilfulness, or by a general lack of contact with the realities of the market-place on the part of an aging and arthritic management.

Melvin-Jones died in 1917. The man he had chosen as his successor was Thomas Findley, the general manager of the Toronto plant. Findley had a lot of qualities that the Masseys, and the company built in their image, found admirable; he was a devout Presbyterian, a conductor of Bible study classes, and the president of the Toronto West End YWCA. In addition, he was able and energetic and a comparatively young man in his forties. All these capabilities and virtues seemed to make him an ideal choice for the job. He was by far the youngest senior executive in the company and the alternative to him would have been a man thirty years older.

However, Findley's leadership was to be tragically short-lived. Within two years he was found to be suffering from a rare and malignant form of cancer. The disease grew worse slowly. But by the middle of 1919, Findley knew he was going to die and, with a heroic sense of responsibility, started casting around for a successor. His search was a hard one since the company was devoid of any young, aggressive executives in its senior ranks. Findley could not risk nominating an older man. He knew such a choice would do nothing to solve the problems that Massey was facing around the world. And so, in some desperation, he turned to an outsider, Thomas Bradshaw, to take up the post of treasurer and manage the company, and he asked Vincent Massey, Chester's elder son, to become president.

* In the U.S. in 1910, there were only 1,000 tractors; by 1920, there were 246,000. A similar expansion took place in the smaller Canadian market.

On a summer evening in 1921, Vincent Massey and his wife, Alice, were driving in the countryside when they were stopped by a farmer. Recognizing him, the farmer addressed Vincent by name: "Mr. Massey, our corn harvester has broken down and we can't find out what the trouble is. Can you help us?" Vincent was a classical scholar, a university professor, and a graduate of Balliol College, Oxford, but as a farm mechanic he had a lot to learn. He could scarcely ignore the request, however, or plead ignorance, so he followed the farmer into the field. There, without knowing what he was doing, Vincent began to tinker with the machine. To his astonishment, the harvester roared into life. The farmer was delighted. And Vincent Massey, soon-to-be president of Massey-Harris, departed with his reputation intact.

The story is one that was told by Vincent Massey himself and it illustrates, with some self-effacement, the degree to which the interests of the younger generation of Masseys had deviated from farm machinery. Vincent had been made a secretary and director of the company and was to be its president for four years. His younger brother, Raymond, also worked for the family firm. But neither brother had his heart in the work. One was to be a famous administrator and diplomat, the other a well-known actor. While others in the new generation of Masseys were to become religious leaders, teachers, artists, doctors, mortgage brokers, and investment dealers, there were none who held any great ambition to go into business or rule over the international empire begun by Hart Massey. The company was coming to a parting of the ways with the Masseys. And the division was to take place at a time of turmoil.

In the early 1920s farmers in Canada received a first bitter taste of the disasters that would strike them ten years later. Immediately after the end of the war, wheat prices had risen in Europe and around the world. Prairie farmers had been hard-pressed to keep up with demand. The price of a bushel of wheat had soared to $2.41. The bubbling prosperity was reflected in farm implement sales and Massey-Harris fared well. The good times, however, did not last long. A surplus of food caused prices to drop. And the fall was a precipitous one. The price of a bushel of wheat went as low as eighty-one cents. Farmers could no longer afford to buy new machinery, nor could they manage to pay off debts on machinery they had purchased only a few years earlier. Massey-Harris was in trouble.

By the fall of 1920, it was evident that the company would lose money. Not only had its Canadian operations been hurt, but losses were turning up at its U.S. plant and in its export markets. In France, sales over the next two years were to be cut to one-third of their previous level. It was in this situation that the terminally ill Findley turned to his lifelong friend, Thomas Bradshaw, for help. Like Vincent Massey, Bradshaw knew virtually nothing about industry or manufacturing, and still less about farm implements. But he was a skilled and renowned financier. A partner in the investment firm of A.E. Ames and Company, he was an expert in municipal finance and had served as a highly successful commissioner of finance for the city of Toronto.

Findley insisted that his friend take the post of treasurer immediately and then become general manager in November. Bradshaw was reluctant. He judged, correctly, that his appointment would be criticized and would meet with a lot of opposition within the company. It was an article of faith at Massey that general managers should be men steeped in the company's products and in manufacturing methods. Never before had a person from outside the industry held such an important post. Bradshaw eventually agreed to accept. But he did so only out of respect and affection for his dying friend.

Findley coupled his desire to get Bradshaw to manage the company with another wish; that Vincent Massey should become president. In the month of his own death, December 1921, Findley was succeeded by the 34-year-old Massey. In the same year the new president had the task of telling the shareholders that the company had lost $1.4 million on sales of $17 million. Losses would have been greater but the financial detective work of Bradshaw had unearthed a war damage claim of $1.5 million against Germany and this was entered as a credit to the company for that year. The only bright piece of news was that a dividend would continue to be paid despite the loss. Not until 1925 could the company afford to be so generous again.

When Vincent Massey reported the company's losses to the annual meeting it was a shattering break with precedent for the company his great-grandfather had founded. Never before in any year since 1847, had the company lost money. What the losses did was to shatter the myth that with its great size and the diversity of its markets, Massey-Harris was immune to recessions and economic troubles. Until now the company had expanded as the process of farm mechanization itself had expanded, and as Canada had grown and the Prairies had been opened up. From now on, there would be less natural expansion. The company would find itself living much closer to the boom and bust of the farm cycle, concerning itself each year with the financial capacity of farmers to borrow money and to buy Massey equipment.

As Massey faced the slump of the early 1920s, there were few signs of immediate relief or of better times to come. Actual management of the company fell to Bradshaw who steered a sound but unimaginative course while he learnt the rudiments of the farm business. The gentlemanly Vincent Massey would drive in from his 400-acre estate at Port Hope to preside over board meetings. But neither of them had the knowledge nor the initiative that was needed to take the company in new directions. It was a case of hanging on and hoping that the recession would run its course and business would pick up.

The wait was a long one. But by 1924 a recovery, which was to improve the company's fortunes, did finally get under way. Meanwhile events were taking place in the personal life of Vincent Massey that were to change the ownership, and the future, of Massey-Harris in a dramatic way.

It is hard to resist the idea that Vincent Massey was a man who found it difficult to fit back into his surroundings. He was an intellectual who moved in an entirely different milieu from the all-too-agrarian directors who made up the board.* He complained about the narrowness of their views, even while he desperately looked around for some useful contribution that he, as president, could make to the company. He advanced a scheme for attracting young men of good liberal education into the company and grooming them for future management positions. The other directors thought nothing of the idea.†

Frustrated with the day-to-day routine, Vincent saw a wider role for Massey and for himself on the international scene. To Vincent, the limited vision of the company's directors and managers was in great contrast to the worldwide business influence of the company they ran. Here, felt Vincent, was a suitable outlet for his energies, and one that fitted with his background and his social ambitions. He would become a kind of ambassador for Massey-Harris, adding to its prestige at home and abroad; thus, in his own mind, at first unconsciously, was born the career that would shape his life, that of being an international diplomat.

---

* Vincent Massey had been educated at the University of Toronto and had taught there; he had also attended Balliol College at Oxford University where he had mixed with Britain's social and intellectual elite. He played a major role in the construction of Hart House at the University of Toronto, and the establishment of the Massey Foundation.

† Massey next turned his attention to public relations. He introduced a new design for the firm's calendar. Three horses were shown pulling a binder over the brow of a hill. Vincent was very proud of it, considering it to be "not great art, but good graphic art." His fellow directors objected strenuously. The horses were pulling too hard, they protested, and that would give farmers the wrong idea about Massey binders.

At first there were trips to Western Canada where he addressed luncheon clubs and farmers' groups. Since the west was at the time beset by a recession which it largely blamed on Eastern Canada, Vincent's presence and his speechmaking helped to soothe some feelings and to ruffle others. His horizons, however, were broader than this. Trips and visits to the branch offices in Europe were planned. These journeys were undertaken with the idea of raising morale, and finding new sales for the company at a depressing time.

Some of Vincent's interventions had an adventurous twist to them. He became interested in the Genoa conference of 1922, an international gathering that was aimed at reviving prerevolutionary patterns of trade with Russia. Massey had been a supplier of farm equipment before the Communist takeover. So Vincent persuaded the Liberal prime minister Mackenzie King to take an active part in the Genoa talks. Concurrently, his own company man, R.O. McDonald — who had headed the Massey office in Moscow — managed to get himself appointed an official member of the British Empire delegation.

In this as in many other things, Mackenzie King proved himself to be indecisive and inconstant, and when the negotiations turned to guaranteeing trade credits, he drew back and ordered that no initiative should be taken. Vincent was dismayed. But he had his own plans to pursue and, in 1924, managed to get an invitation to Russia. Crossing the frontier under a revolutionary banner that read "Welcome to the Proletarians of the World," Vincent and Alice spent two weeks touring the wrecked and dilapidated cities of Moscow and Leningrad which had been torn apart by the fighting of the civil war. In Leningrad they found the streets deserted, houses stripped completely for firewood, and the city on the verge of ruin. Plainly no business was to be conducted, but the visit, followed by a tour of Poland, Austria, Hungary, and Czechoslovakia, had been of value to Vincent personally.

Trips abroad were matched by an ambassadorial approach to lobbying the government in Ottawa. Vincent quickly became disillusioned with the Liberal government of Mackenzie King, which was dependent for its survival on the support of sixty-five Progressive members from Western Canada. These members were pressing for a further lowering of tariffs on farm machinery. And Vincent had to oppose them. In meetings with ministers and with Mackenzie King, he argued that such a policy would damage the manufacturing industry in Canada. American imports would throw hundreds of Ontario workers out of jobs. Vincent also wrote a letter to the Conservative opposition leader, Arthur Meighen, protesting his party's position on freer trade.

The Liberal budget that was drawn up that year did in fact lower tariffs. Vincent wrote another letter to Meighen in which he proclaimed his disgust with the government's actions: "The whole attitude of the cabinet, in this matter, seems to reveal an even mixture of cynicism and hypocrisy." It was an unwise letter. For, within a year, Vicent's political ambitions were to lead him into standing as a Liberal election candidate. The letter he had sent to the Opposition leader suggested that, if ever there was a case of political cynicism and hypocrisy, it could be found in Vincent Massey's dramatic conversion to the Liberal cause. Moreover, even before the election, he had been offered — and had accepted — a cabinet post as minister without portfolio in Mackenzie King's government.

Having handed his opponents a weapon with which they could mortally wound him, Vincent Massey proceeded to turn down offers of safe Liberal ridings in Stormont and northeast Toronto and decided, instead, to contest Durham, the constituency in which his country home was situated. This was to his credit since he had no wish to use his influence to get into Parliament. But it betrayed a total lack of the realities of politics. Durham had elected a Conservative in every election for the previous twenty-one years.

Vincent's campaign was doomed from the start. It was not helped along by his position as a newcomer to the community, his status as a landowning man of wealth, and his habit of having his chauffeur drive him to political meetings (When he was urged to drive himself, Vincent refused on the grounds that it would be dishonest since he was normally driven by a chauffeur).* The letter he had written to Meighen also hung over the campaign. The Conservative leader threatened to reveal its contents. Vincent argued lamely that this would be a breach of confidence since the letter had been marked private. In any event, revelations were unnecessary. On polling day, the result of the voting was Fred Brown for the Conservatives 7,020 votes, Vincent Massey for the Liberals 6,074.

Disappointed in politics but now committed to public life, Vincent Massey's career was to take off in a different direction with his appointment in 1927 as Canada's first minister to the United States. He was subsequently posted to London in 1935 as High Commissioner and eventually became Canada's first Canadian-born Governor-General. Throughout his years of public service, it is impossible not to think that the sense of duty and the deep feeling of responsibility for others that had animated his family through three generations did not burn with similar intensity in Vincent. Cultured, intelligent, aristocratic, almost

---

* After sitting through one of Vincent's rallies a loyal Liberal sent a bill to the party's agent. He apparently expected to be paid for his attendance.

the epitome of the English country squire, Vincent Massey managed at the same time to have much in common with his stiff-principled forebears, Daniel and Hart.

Vincent's initial sally into politics was to prove a turning point for Massey-Harris. At the time it was merely a source of conflict and confusion. But the ill-fated Durham election marked the beginning of the ultimate separation between the Massey family and the company — the end of a family dynasty and of an era.

Rumors and conflicts about the intentions of the Massey family had started with Vincent's trips to Ottawa to lobby on tariffs. As the argument over tariffs raged within the country, chiefly in the form of a division between west and east, the spectacle of a prominent Ontario industrialist being received by members of the cabinet and the prime minister revived old suspicions. The Masseys were influence-peddling. Worse than that, they were engaged in a game of cat and mouse, black-mailing the government to get their way. The result, it was feared, was that the tariff would stay and the west would lose out again. The Lethbridge *Herald* on March 17, 1924, wrote: "It is stated on excellent authority that Mr. Massey has threatened to sell out his plant lock, stock and barrel, and bring to a close the career of the lone implement concern in Canada. It has been known for some years that the International Harvester Company of the United States has been ready to acquire Massey-Harris... and thus eliminate all competition in the implement business. Mr. Massey, according to a reliable source of information, has now decided that if the Government makes a cut in the tariff on these implements of any consequence he will throw up his hands, sell out at good advantage to himself, and close his factory."

Vincent Massey had no such intentions. Nor would he present an ultimatum along these lines unless he planned to carry it out. Deviousness was not a part of his nature. Nonetheless, right or wrong, the idea had taken root that the Massey family might be party to a sellout to American interests.

In the summer of 1925, as Vincent's political plans began to take shape, his position at Massey-Harris rapidly became untenable. When he first informed the board of directors that he had been asked to stand as a Liberal candidate and become a minister, they accepted this and gave him permission to go ahead with his plans while remaining company president. The decision was made because Vincent, through himself and his position as a trustee of the Massey Foundation, held a controlling interest in the company and various board members did not want to upset him. However, the decision was taken when the two strongest board members were absent; Lloyd Harris, son

of the founder of the Harris Company, and a staunch Tory,
and Walter Massey's wife, Susan, Vincent's aunt.

When they heard about the decision, they forced another meeting of the board at which they voiced strenuous objection. Although Vincent still had some defenders, it was plain the prevailing feeling was that politics and business should not be mixed. Moreover, there was among the directors — and among the country at large — a great deal of antipathy to the Mackenzie King government. It appeared that Vincent had not chosen the right moment to get involved in politics. At the close of the meeting he agreed to resign the presidency and his position on the board of directors. He also stood down from two other directorships, with the Canadian Bank of Commerce and the Mutual Life Assurance Company. When he fought his electoral battle in Durham, Vincent did so as a private citizen. The job of running Massey-Harris was being carried out by Thomas Bradshaw.

Rumors, however, continued to circulate about a U.S. takeover of Massey-Harris. And these did begin to have some substance in fact. Following the failure to develop its own tractor and get into the industry's newest and fastest-growing market, the firm had been casting around for another company to link up with. Under Bradshaw's direction, it had settled on the J.I. Case Plow Works Company of Racine, Wisconsin. Massey's intention was to obtain the rights to the Wallis tractor, manufactured by Case, and to market it.

The negotiations took a different turn. Aware of Vincent Massey's position, the Americans went on the offensive and the directors of Case approached Massey-Harris management in a bid to gain control. Case would buy the shares held by the Massey Foundation and the Massey family. As it happened, the offer came at a critical moment. Vincent had been approached about accepting the ministerial post in Washington. He was concerned that there would be a conflict of interest. His new position would involve him in negotiations over tariffs and farm policies; how could he continue to be the largest shareholder in Canada's most powerful farm machinery company? Was the time not ripe for him to end his association with the firm and pursue his diplomatic career instead?

The affairs of Massey-Harris had been improving. However, speculation about the company's future exaggerated any improvement that had been made. The stock exchange was swept with rumors of a bountiful offer being made to all shareholders and — since stock market activity in the twenties was hectic anyway — the share price went higher and higher. For shareholders in Massey the feverish trading and high price of the stock was good news. But for the managers of the company, the prospects of a buy-out by Americans was hard to accept. When Case's offer was made known, Thomas Bradshaw had been on a train going west, on the first part of a trip to Australia. At Port Arthur, (Thunder Bay), he got off and headed back to Toronto.

A widely-held Canadian-owned company.

Bradshaw reasoned that Vincent Massey did not want to deal with the Americans and would prefer to keep the company in Canadian hands. However, it was plain that Vincent was determined to sell. The Washington appointment seemed to represent too much of a conflict. Obviously if no other offer presented itself, Vincent Massey would have little alternative and Massey-Harris would become the branch plant of a large American company. Bradshaw, with his expertise in financing, knew that he had to move quickly and effectively. Within a matter of weeks, he would have to organize a Canadian syndicate that could make a good enough offer to Vincent. To buy his 70,000 shares, Bradshaw needed $8 million.

Back in Toronto, one of Bradshaw's first calls was at 36 King Street West, the offices of Wood Gundy Limited. Bradshaw and Harry Gundy were old friends and colleagues from the war years when Gundy had been the most vigorous member of the National War Finance Committee, and had helped organize war loan drives which had made one million Canadians purchasers of bonds and supporters of the war effort. James Henry Gundy was now forty-seven years old, but he had lost none of his verve or vitality. The son of a Methodist minister and a comparatively poor man, he had joined up with George Herbert Wood, thirteen years his senior and a man of impeccable social credentials, to found the Wood Gundy partnership in 1905. For several years afterwards, most of the firm's business was in the sale of government bonds — an association which had brought Gundy in close contact with Thomas Bradshaw, the municipal financier.

Gundy was a superb salesman, an enthusiast who could persuade others of the merits of any scheme; he was also a popularizer, keen on attracting new business from the resource industries that were starting up and looking for capital. He saw a natural match between the great mining concerns that were developing mineral riches in remote places like Noranda and Sudbury, and the stock market investors' desire to buy in and be part of Canada's first concerted resource boom.

Twenty years before, when Wood and Gundy had

combined forces, nearly all investment activity had been in government and municipal financing. It had been an exclusive and clubby business in which the public at large did not participate. Investment in the bond market was for the wealthy. Gundy, with his poor and nonestablishment background, had wanted to change this. As a salesman it was a situation that frustrated him, and as a businessman with a shrewd eye for a good investment, he saw the potential of strong active markets for stocks and bonds. Five years before Gundy was approached about Massey-Harris, Wood Gundy had undertaken its first corporate underwriting, a striking departure from the past and a risky thing to do. It had been a success and the firm had followed up with new issues for several big Canadian companies, putting itself in a leadership position in the field.

When Bradshaw talked about the need to buy out Vincent Massey, Gundy listened carefully. A lot of money was involved but it was the kind of project that appealed to him. No one could doubt the solid financial base of the Massey-Harris company or the great opportunity that purchase of a controlling shareholding would offer at a time when the stock market was roaring ahead. Moreover, the public and the government would applaud the deal. Gundy and Bradshaw could cast themselves as the white knights of Canadian nationalism fending off the threat of an American takeover and saving a grand old manufacturing firm.

The scheme Gundy devised was an attractive one. He and a syndicate of investors would purchase the Massey interests. Subsequently the $100 per value shares would be split into four with no par value and this would ensure a wide distribution of the stock among Canadian shareholders. At the same time the company would improve its financial position with a $12 million bond issue that would allow it to pay off outstanding bank loans. The net effect of this would be a healthier balance sheet and the transformation of Massey-Harris into a widely-held but Canadian-owned and managed company. The deal would confirm Gundy as the senior member of the board.

When the proposal was put before Vincent Massey, he accepted it. Before the sale could be finalized, however, there was to be a clash of wills between the old owner and the new ones, and for a time it looked as though the deal might fall through. In their desire to glamorize what they were doing and court public opinion, Bradshaw and Gundy had let it be known that if they did not succeed in buying Massey-Harris the Americans would. This was something of an exaggeration. While an offer had been made, Vincent had not indicated that he would actually accept it, merely that he saw a conflict of interest and wanted to be rid of his position in the firm.

The newspapers, with their old suspicions of the Masseys, had taken up the chorus and were busy denouncing Vincent while praising Bradshaw and Gundy.

Perhaps emboldened by this, a statement about the sale was drawn up for the newspapers which referred to the patriotic action of the two. When Vincent Massey read it, he was furious and adamantly refused to approve the sale. The deal was supposed to be completed in the offices of the National Trust Company, which had conducted the negotiations on behalf of Vincent; after the signing, the new Canadian minister to Washington had to catch a train to take up his post. For a time Bradshaw and Gundy would not back down. Nor would Vincent Massey. His wife caught the scheduled train but Vincent would not depart before the matter was resolved. In the offices of the trust company the arguments and wrangling continued.

Vincent drafted a new press release. It spoke of the Masseys having "a keen appreciation of the desirability to Canada and to Canadians of retaining to this country this important institution with its worldwide trade and its opportunities for greater extension and progress." Bradshaw was asked to sign it but refused. He agreed that it was a correct statement but said he could not approve it because Vincent was forcing the issue and holding a pistol to his head. The stalemate continued for a time. Eventually, after discussion with other members of the syndicate, Bradshaw reluctantly initialled the document. Vincent Massey then departed for Washington, and in doing so ended the family association with the firm. It was the spring of 1927. And it was exactly eighty years since Daniel Massey, with boyish enthusiasm, had bought out a bankrupt foundry in Newcastle, Ontario, and had begun to repair and manufacture implements for local farmers.

Part Two:
The Duncan Years (1927-1956)

5

Winners and Losers

The years that lay ahead, the 1930s, were going to be a testing time for the company during which its survival would be in doubt. As a perceptive and experienced farm machinery man, Duncan almost certainly sensed this ahead of time; he could see how political and economic events were unfolding in Europe.

One of the legacies of any great company is the sense of loyalty it engenders and the dynasties that are often fostered within it. For Massey-Harris, the succession of able and vigorous members of the Massey family — who created the company and then led it — left the greatest imprint, through Daniel and Hart to Charles and Walter. That legacy came to an end with Vincent's decision to sell out and pursue a diplomatic career. In the aftermath of the buy-in by Harry Gundy's syndicate, ownership of the firm was widely held. Such diversity did not of itself present a problem, at least for the time being. But it did signify two things that were to be of great importance later; first, that since ownership and management were no longer in the hands of a single group there was the possibility for conflict where none had existed in the past, second, that new opportunities now existed for outsiders to acquire company stock and, if they purchased enough of it, exercise control.

The takeover by professional investors led by Gundy had repercussions for the way Massey-Harris was to be run. If the company's performance did not meet investors' expectations they might get restive, particularly if dividends had to be cut or go unpaid. They could possibly get angry enough to dismiss one manager and appoint another. But there would be no personal involvement by shareholders in management as there had been under the Massey family. The men chosen to run the company would have more authority and power.

It was in the management of the company that Massey-Harris nurtured another dynasty, whose source was unusual. James Duncan, the manager of the firm's operations in France, had a son, also called James Duncan, who was to rise to the top and run the company for twenty turbulent years. Duncan was to become president in a storm of controversy, and at a time when Massey-Harris was at the brink of bankruptcy. He was to depart the company amid another controversy. In between he would pilot the firm through the war years and through a great period of international expansion.

All this was far in the future in 1907 when the younger James Duncan, then aged fourteen, received the first intimations of what could be ahead. It was in his father's home at Gagny in France and the visitor that evening was the autocratic Lyman Melvin-Jones. The senator and president of the company enjoyed having an attentive audience and he had found a good listener in young James. After dinner, as he smoked a cigar with his host, Melvin-Jones grew expansive and began recounting stories of his days in the Canadian west, and of the excitement and adventure of pioneering in the Red River valley. James, who had never visited Canada, was fascinated. To a young boy used to the disciplines of Paris and its foreign enclave, the Canadian west conjured up new and romantic vistas. And Melvin-Jones, flattered by his attention, led him towards these. Turning to the elder Duncan, he observed that young Jimmy would make "a real implement man" and, in due course, should be sent to Canada to learn the business.

Three years later James Duncan was to get his first glimpse of Canada. As his train crossed the American border en route for Toronto, it was not the endless horizons of the prairies he found himself gazing at, but the wintry landscape of Niagara. Once he arrived at the city he was to stay for two nights at the Walker House Hotel before finding humbler lodgings. Two days later he was to report to the King Street factory. There, the president of the company had promised his father that he would be given a job at a salary of $9 a week. It was not the romantic dream that James had imagined. He was told he would be an assistant in the seeding machines department.

At the age of seventeen, James Duncan was a mature and worldly young man. It was, he noted later, a foregone conclusion that he would follow in his father's footsteps, and his young life had been modelled around this. Conventional schooling had been done away with. Instead he had been tutored in subjects that would prepare him for a business career, and in foreign languages. His holidays, on Thursday and Sunday mornings, had been spent assembling farm machinery. Parts

were sent to France from Toronto to be assembled there.
Mechanics to maintain the machines and demonstrate them to customers were also sent to France. Few of these mechanics spoke French, and so acting as an interpreter was another task that James had had to perform.

Hundreds of farmers and dealers would turn up to see a Massey-Harris field trial. First, the Canadians would demonstrate a machine, then young James would address the gathering, reading from a prepared text or translating into French the words of the English-speaking mechanics. At fifteen, his father would occasionally send James to do the demonstrations himself and book the orders.

Early responsibility had made James an unusually self-reliant and confident person for his age. He also had another quality that was to mark him for the future; as a young man of good family from Europe, he had no difficulty in impressing his superiors not only with his abilities, but also with his social maturity. Perhaps it was the provincialism of Canadian society or the fact that (as Vincent Massey later pointed out to the directors) few young men of good social standing and liberal education were drawn into the farm machinery business, at any rate James Duncan was singled out, almost from the beginning, as a competent and hardworking young man, who also had the right social background. Quick to acquire patrons in the company, he was destined to rise fast.

Like any other ambitious young person, James encouraged this process along. He had arrived in Toronto with a number of advantages, not least the imprimatur of the president, and the reputation that he carried with him as his father's son.

He went to work first in the plant where his fellow workers ridiculed his English accent and offended him by chewing tobacco and spitting on the floor. Then he moved into sales, demonstrating machinery for farmers in Ontario as he had done in France, and travelling across the province by rail in summer and in winter by sleigh. As a salesman, his salary was raised to $14 a week. At the end of the financial year when the company closed its books, the young salesmen had to work overtime to settle up their accounts. By custom, they had always been paid extra for this additional work. James refused the extra remuneration. The company's interests came first to him and he did not want to be paid. His colleagues were angry, but he could not be persuaded to change his mind.

James was in fact leading two lives — a working one and a social one — on two different levels. The social side of his life made him as familiar with the bosses, Melvin-Jones and Thomas Findley, the directors of the firm and other luminaries of Toronto society, as he was with his everyday colleagues. In

Paris his family had not been regular attenders at church. But in Toronto James soon recognized that it was de rigueur to be a churchgoer, particularly at Massey-Harris. So within a matter of months, his Sunday holiday was taken over by church services and Bible classes.

In the morning he would attend the service at Bloor Street Presbyterian Church patronized by the general manager, Thomas Findley, and afterwards return with the Findley family for lunch at their home. In the afternoon he would be an attentive listener at Findley's Bible class, then, after having tea in the home of Joseph Shenstone or one of the other Massey directors, go with them to evening church services. On alternate Sundays he would go to dinner with the president; the two of them would dine alone at Melvin-Jones' St. George Street home. Melvin-Jones had only one daughter who was married to the minister of St. Andrew's Presbyterian Church, and as with many egocentric personalities he was a lonely man. Over a bottle of rare French wine, Lyman Melvin-Jones would reminisce about the past and about his own life.

Plainly James Duncan was marked for success. He might have made swifter progress in the next few years but for an interruption which was to set back all young men of his generation, the First World War. The German invasion of France created a host of problems for his father. Massey-Harris's staff were called into the armed forces and the French government declared a moratorium on all debts. The older Duncan sent a telegram to Melvin-Jones in Toronto requesting his son's return to Paris to help him.

For the next two years both Duncans, father and son, were kept busy trying to maintain Massey-Harris as a going concern. James, now twenty-one years old, bought a Model T Ford in which he travelled around the French countryside attempting to persuade dealers to pay their debts to the company. This was tricky since, under the debt moratorium, they were not obliged to pay up. He would also travel to the ports and organize the unloading and transportation of farm machinery shipped in from Canada and England.

As the fighting on the western front intensified, the younger James came more and more to the realization that he must join the army and go to war himself.* In July, 1919, he reported to the British recruiting officer in Paris. For the next few years

* A military exploit was to precede Duncan's entry into the armed forces. By agreement with the French ministry of supply, he supervised the evacuation of a warehouse full of Massey-Harris farm machinery that was within sight and shelling distance of the German lines. The machines were taken out under the cover of darkness by Duncan and a party of French soldiers. They were needed to raise farm production in the western and northern regions of the country to which refugees were returning. The exploit was later the subject of an oil painting which now hangs in the Argus offices of financier Conrad Black.

James Duncan was to see plenty of action as he fought with the British army in France, rising to the rank of captain in the Royal Field Artillery. He was not demobilized until March, 1919. Only then was he able to resume his career with the firm, and he did so in typically businesslike fashion. Many twenty-six-year old officers, who had gone through the terrible years on the western front, would have stayed on in London to celebrate their demobilization and their new-found freedom. Not James Duncan. He immediately journeyed to Paris and the following day was back at work again.

The return to Massey-Harris in France satisfied Duncan for a time. But it did not fulfill the ambitions that he had nourished since his prewar days in Canada. And so, several months later, he sent messages to Toronto indicating that he would like a job with greater responsibility, and that he felt this was his proper recompense. Management in Toronto was not used to getting such ultimatums. But Duncan had powerful friends and his wish was granted. He was put in charge of the company's operations in southern Europe and given the task of building sales in Italy and Spain as well as the countries of North Africa. As it turned out, the appointment was a good one. During the war years, these countries had run short of farm equipment. Now they began importing large quantities of machinery in an effort to raise food production. Duncan was able to increase sales tremendously in a short period of time.

At head office in Canada, the early 1920s were a period of management transition, of poor results, and financial losses. They were also the period when the young Vincent Massey began influencing the company's international operations, and talked of the need to achieve more growth abroad, and to attract a new breed of well-educated, young executives. In Vincent Massey, Duncan had a natural ally and a powerful patron. The president told the directors and management that James Duncan was the most capable young manager they had and should be promoted. In return he was informed that, for a man of thirty, Duncan was already doing well, perhaps too well.

In 1922, he was sent to Mexico at Vincent Massey's urging to investigate the possibility of opening a sales office. From there he was dispatched to the Balkans and the Middle East. Following the death of his father, he was made acting manager for France, Holland, Belgium, and North Africa. In 1925, he reported to the head office on the advisability of establishing a manufacturing plant in France.* Subsequently it was decided

---

* Duncan's father had advised the establishment of a manufacturing plant in France before the war, but his advice was ignored. It became more urgent when, in 1924, the International Harvester Company of Chicago started a factory in France and began taking business away from Massey.

that Massey-Harris would set up a factory, and a location was chosen near Lille. It was the first Massey-Harris factory outside North America. Much of the credit for this pioneering move went to Duncan, and six months after his father's death he was confirmed in the position that the senior Duncan had held.

James Duncan was moving ahead fast. He was doing so on the basis of his own ability, and because of the long-standing relationships he had built up with the senior men in Toronto before the war. Findley, whose Bible class he had attended, had run the company briefly. He was succeeded by Vincent Massey, who thought highly enough of Duncan to later offer him a job as first secretary on his staff in Washington. (Duncan politely declined the offer.) Following Vincent's departure, the office of president was filled by Joseph Shenstone, again an old contact of Duncan from his Toronto days.

The only person Duncan did not know well was Thomas Bradshaw who, after 1927, was effectively running the company. Bradshaw had less of an interest in the European operations, and less enthusiasm for Duncan. While he realized that the young European manager knew the farm business exceedingly well and was an industrious, capable salesman, he had doubts about his youth and his ability to head the company's manufacturing and engineering side as well as sales and finance. Duncan had been given the title of European general manager, but Bradshaw cut the job in half and sent his executive assistant, Cecil Milne, to Europe with the same seniority as Duncan to take charge of manufacturing and engineering. Duncan was prepared to work with Milne, though he was disappointed by the decision. It was a setback, but not one that was irreparable for an ambitious young man who had achieved much and was still in his mid-thirties.

The decade that lay ahead, the 1930s, was going to be a testing time for the company during which its very survival would be in doubt. As a perceptive and experienced farm machinery man, Duncan almost certainly sensed this ahead of time; he could see how political and economic events were unfolding in Europe, and the problems this could create for Massey.

Manufacturing abroad.

In the mid-1920s the world was becoming a smaller place. Hart Massey, in his original enthusiasm for doing business around the world, had envisioned a global market for his products. And this was a reality that seemed achievable as the company set up agencies and distributorships outside Canada, and began boasting of its sales in far-off places. But the twenties were to be a great disappointment for Massey-Harris. In the wake of a great war, nations were plagued by internal problems. In a bid to keep their own costs below those of neighboring countries and to make home industries more competitive, nations everywhere put up tariffs, cut back on foreign trade, and resorted to currency devaluation.

For an international company like Massey-Harris, the pursuit of these policies created an immediate and alarming crisis. The company had prospered by exporting from its vast factory in Toronto and other Canadian locations, and because Massey's operations were so large and so centralized it could sell its equipment competitively. Now, all of a sudden, there were a hundred reasons why it could no longer price its goods lower. There were tariffs and import duties. There was the cost of freight and packing. There were currency devaluations and customs inspections and endless forms to fill in and bureaucratic problems to resolve. All these hurdles were intentional for in the markets that Massey was selling into, governments had decided to protect and favor their own industries and keep foreign-made equipment out.

The logical solution was for Massey — alone among Canadian companies in facing these global problems — to become a local manufacturer instead of an international one. This was the policy that Thomas Bradshaw arrived at, and pursued, though he realized it was going to be costly. Manufacturing abroad would mean less work at home, which in turn meant idle Canadian plants and Canadian workers. Nonetheless, the situation was grave; in France, the company's largest foreign market, sales were cut by two-thirds in the early 1920s. Bradshaw knew he had little choice. If the company was to survive at all in France, Massey-Harris had to become a French firm, manufacturing the latest farm equipment locally. And the same applied in other important markets; in Germany, Australia, and the U.S.

Slowly and methodically, over the next five years, Bradshaw was to direct Massey into a new path. He was not an imaginative leader, nor did he pretend to have any mastery of manufacturing or the farm machinery business, but he was an accomplished man of finance, and logic decreed that the company's relationships with its overseas markets had to change, even if this was not always immediately for the better. First there was the factory in France. Situated at Marquette les Lille in northern France, the factory was within sight of International Harvester's plant. It was close to the canal routes into Belgium, and it could be expanded to manufacture binders as well as haying equipment if the French government should add yet another tariff. The factory was administered by a brand new French company, Cie Massey-Harris S.A. In Germany, an unused war factory which had been left vacant by a truck manufacturer was bought at Westhoven, near Cologne. Unlike the French factory, it did not prove too much of a success. Quality was poor, and the German farm economy was in a state of disarray. Still, by making a move into Germany as well as France, Massey had established a larger presence in continental Europe, and this was to be of significance later.

It had also done the same thing in Australia. Following the sudden imposition of tough import laws there in 1930, a team from Toronto opened negotiations with the H.V. McKay Company, Australia's largest manufacturer of farm equipment, about merging their two operations. Eventually this was done. But with the Australian government threatening to drive out all foreign suppliers, including Massey, it was a case of trying to survive and there was little hope of striking a good deal. Massey ended up surrendering all its Australian assets and goodwill for a twenty-six percent interest in McKay.

In the U.S., the situation was similar — Massey had to get into manufacturing. A certain amount of havoc had been created by the J.I. Case Company which had turned Massey's interest in obtaining the sales rights to its Wallis tractor into a bid for Vincent Massey's shareholding. Although by joining forces Gundy and Bradshaw had succeeded in resisting the American bid, they still faced the problem which had started the tussle in the first place — Massey did not have a satisfactory tractor.

To rectify this Bradshaw went back to Case and successfully negotiated a marketing agreement for the Wallis tractor. Next, with help from Harry Gundy, he returned the compliment that Case had earlier tried to pay to Massey; in 1928, Case was purchased for $2.4 million. Although the owners were allowed to retain the name and to found a new Case company, Massey acquired their plant in Racine, Wisconsin, to go along with the Johnston plant it already owned in Batavia. Expansion seemed to be under way in the U.S., the world's largest farm market. And, perhaps most important of all, Massey-Harris finally had a good tractor.

All this appeared to be a great beginning. In France, production was forging ahead and sales were rising. In the U.S. sales offices were added and the Wallis tractor line started bringing in new business. By 1929, sales in the U.S. were outpacing those in all other markets; $14.8 million against $13.5 million in Canada and $5 million in France.

Not unnaturally, to the men who ran Massey-Harris,

seemed as though the bad times of the mid-1920s were over. The right decisions had been made, as was evident by the orders flowing into the company's old and new factories around the world. Their optimism was shared by others. In 1928, the world's farmers had produced a huge grain crop, breaking all previous records. On the Prairies Canadian wheat pools had been left with a wheat surplus that they could not sell of 100 million bushels. Far from bemoaning this oversupply, the farm organizations held back confident that the crop next year would be average and they would be able to sell into a rising market and get better prices.

Within Massey, the confidence of the farm community was applauded. Keeping the surplus grain in storage rather than accepting low prices for it was beneficial for the firm. It would mean a bonanza in farm equipment sales in 1929 and 1930 after the farmers had enriched themselves. And production was geared up accordingly.

The bonanza did not come. The calculated decision to keep Canadian wheat back from the market proved to be one of the worst mistakes ever made. And when the stock market crashed on Black Tuesday, October 29, 1929, it marked the onset of the Great Depression. The unsettling conditions of the middle- and late-1920s — the high tariffs and quotas and cartels and subsidies and other obstacles to trade — had been the prelude to something much larger; the whole world had been hurt by them, and by the riotous speculation in the stock markets, and now each individual country would pay the price.

The Great Depression.

For Canada the price was going to be a heavy one. And it would take its toll on the farm industry. By the summer of 1929, the wheat pools were getting uneasy. There had been huge crops in Australia and Argentina and there were prospects of another bumper harvest in North America. Prices for wheat had hit $1.60 a bushel in July but, from then on, they began to fall. When the Canadians had stayed away from the market in 1928, the Argentinians had stepped in and taken away their customers in Europe.* By the end of 1929, the price of wheat had been cut in half, and the Canadians were no longer in a

* The Argentinians and Australians were in a position to benefit from any holdback by Canadian farmers since their wheat crop was sown from April to August and harvested from November to January.

position to sell their surplus even if they had wanted to. New tariffs had been slapped on wheat imports by the European countries anxious to protect their own farmers. The price of wheat was now 70 cents a bushel; in Germany it carried an import duty of $1.62 a bushel.

For prairie farmers, this meant a devestating loss. While wheat worth $200 million lay stored in the grain elevators, they were forced to appeal to the government for help. As cold Christmas weather swept across the west, families were given relief money in the form of an advance which they would have to pay back later. With their livelihood taken away by the grain fiasco, relief money was the only way westerners could survive until the spring. Small businesses facing ruin also needed help. And the big companies serving the west — the rail and shipping companies and the farm equipment makers — suffered crippling losses.

The misery of the Depression was to last through the decade. It had been ushered in by the great crash on Wall Street. The desire of stock-hungry speculators to rid themselves of everything they had purchased had gripped the Canadian stock market too. There were huge declines in stock prices, including Massey-Harris stock; from the high of 1929 to the low of 1932, some $6 billion was wiped off the value of blue-chip Canadian companies. There were frauds and allegations of scandals. And the financial community — which three years ago had banded together to purchase control of Massey — was indicted. Sixteen senior partners in the country's top brokerage houses were sent to jail or fined on fraud charges.

The spectacle of stock promoters and manipulators being pilloried and wealthy stock buyers being reduced to poverty did not disturb the public conscience too much. But this was to be only the first chapter of the Great Depression. Far worse was to follow. By the middle of 1930 about 400,000 Canadians were out of work. And among those who had jobs, hundreds of thousands were living on minimum wages.

Dust-bowl conditions gripped the Prairies. The price of wheat had dropped to a mere 25 cents a bushel, less than it would cost to harvest.

The natural deep-rooted grasses of the plains had, since settlement began in the 1880s, been plowed under to raise wheat and corn or turned over to cattle for grazing. Left bare, the soil was dried out by the winds of the Midwest, and soon the rich topsoil was blown away. In the spring of 1932, farmlands from Texas to Saskatchewan were hit by a late freeze, followed by violent storms, a plague of insects, and then a searing drought.

The first great dust storms or "black blizzards" began in

late 1933. Vast clouds of dust particles were carried up into the
atmosphere blocking out the sun for four days at a time. Dirt
blew under windowsills and into homes; settling on food and
drinking water and machinery parts. Farm families stuffed
rags into windows and wall joints and hung wet blankets over
the doorway in a futile attempt to keep it out. Drought
conditions persisted for a further two years, with the storms
spreading north and eastward. Lake Michigan and Lake Huron
dropped to their lowest recorded levels. Towards the end of
1934, intense storms darkened the skies over the Ohio River
valley and along the Great Lakes. In the lands worst affected
by the drought, the so-called dust bowl, crops came out of
the ground stunted and brown; livestock died of starvation, or
sometimes suffocated; people died of respiratory ailments; and
thousands of families packed up their belongings and abandoned
their farms.

Year after year, the deprivation and the sense of hopeless-
ness went on. The government, first under Mackenzie King and
then under R.B. Bennett and the Tories, seemed powerless to
do anything. Government welfare, of a sort, was offered, but
"relief" as it was called, was inadequate. In the rural areas, life
was impossibly harsh. A family of five in Saskatchewan would
be given $10 a month and a sack of flour. None of the money
could be spent on anything apart from dried beans and potatoes
because nothing else was available. Rations for relief, commented
Lewis Stubbs, a Manitoba judge, were not enough to live on
and not enough to die on. By 1933, the average income on the
Prairies had dropped to one-fourth of what it had been four
years earlier.

The ruin and devastation were felt by every business in the
land. But in the farm machinery business the situation was
disastrous. It was not simply a question of accepting the reality
that a ruined farm industry could not afford to invest in new
equipment and that losses would continue to mount, there
was also a sense of demoralization and of helplessness.*

By the summer of 1930, Massey was cutting back in a
desperate attempt to survive. Over half the workers in its North
American factories were laid off; office staff was reduced; and
the managers in Toronto began to wonder if they could
sustain the overseas factories and sales branches. The economic
malaise was worldwide, and the prospects of recovery seemed
bleak. From a profit position of $2.8 million in 1929, Massey

---

* Years later, a Massey personnel manager would recall that when he was given the job
of recruiting young and talented graduates into the company in the late 1940s and
early 1950s, his appointment was opposed by the senior men in the company who
had vivid memories of the 1930s. What was the point or purpose, they wanted to
know, in attracting young men and planning for the future in such a precarious
industry?

recorded a loss of $2.3 million in 1930.

The bearer of this unwelcome news to the shareholders was Thomas Bradshaw. A few years earlier, Bradshaw had allied himself with Harry Gundy, and together they had engineered the acquisition of Vincent Massey's stock, but in this moment of crisis, with sales and profits plummeting, Bradshaw began to learn that friendships are often stronger in good times than in bad.

The Massey board of directors under Gundy's leadership consisted of several prominent members of the financial community whose reputations had been brushed by scandal in the stock market frauds of the previous year. Less worried about their reputations than about their reduced wealth, they rallied behind Harry Gundy in demanding that, despite the parlous condition of its finances, Massey should pay dividends on common and preferred stock. For Bradshaw, this went against the sound principles of financial management that had always guided him. It was not just that the company had sustained a loss that year. Looking ahead, and realizing how desperate the outlook was for the farm industry, he could not foresee any kind of improvement. In his view it was necessary to retrench and cut back.

The argument between Gundy and Bradshaw raged on for several days. It was a confrontation between a manager who felt his responsibilities to the company made it necessary for him to resist and the demands of an ownership group who felt that the company had obligations that it must honor. Such a scene was to be replayed at Massey in later years. In this case, the deck was stacked against Bradshaw. Rattled and nervous about what was happening, Gundy and his fellow directors were looking for someone to blame for the losses. In September of 1930, Bradshaw's resignation was accepted by the board.

At the brink — for the first time.

For the next five years, Massey-Harris was to suffer annual losses. The worst year was 1931 when the losses totalled $4 million. By the end of the financial period in 1935, the company had accumulated a deficit on capital account of $22 million. This was a huge sum for the time, and it put the survival of the company, and its international operations, in jeopardy. There seemed little likelihood of a reversal of fortunes. The U.S. manufacturing company was the most burdensome, racking up one loss-making year after another.* Canadian sales were running at about one-quarter of their pre-Depression levels. The Argentinian and Australian markets had slumped. Only in Europe was business sustained reasonably well for a time. But the plight of Massey was so serious that in all the countries in which it operated, whether they were in the midst of a slump or enjoying marginal prosperity, management was forced to cut costs and prepare for the worst. By the middle of the 1930s, Massey was insolvent and at the brink of bankruptcy.

The dreadful hemorrhaging and the losses were a vindication of Thomas Bradshaw and the position he had taken. The company could not afford to pay dividends. Nor, as the years went by, did it seem likely that it could afford the cost of staying in business.

Massey needed an infusion of new bank loans, and it needed time to pay off arrears of interest and to meet the obligations of servicing its debt. Fortunately, time was granted. The bank which had established its business with the firm in the days of Hart Massey, the Canadian Bank of Commerce, continued to extend new loans in the worst years of the 1930s. And rather than drive the company into bankruptcy, its creditors and bondholders held off and did not press their claims. Many businesses faced the same predicament as Massey, though few were in as desperate a situation. To have foreclosed on Massey would have been to strike a blow at the whole farm-based Canadian economy, and to call into question the ability of any manufacturing enterprise to survive.

The loans and the support the company received were conditional on drastic action being taken to trim unnecessary expenses, get production into line with sales, and establish a healthier and smaller operating company. When Bradshaw resigned, another financier was chosen to led the company, T.A. Russell, but as acting president only. Russell had been a director since 1924, but he had no knowledge of the farm machinery business. What he did have, and what was to be a priceless asset in the years of the Depression, was the confidence of the Bank of Commerce. A member of the board of directors at the bank, he would champion Massey interests to his fellow directors, persuading them of the importance and the significance the company had for thousands of Canadians.

Massey could not be allowed to perish, Russell contended. At the same time, and this he would admit, the firm could not expect open-handed support. Sacrifices would have to be

---

* In 1929, with its new acquisitions, Massey had made a promising start in the U.S. by pushing sales to $14.8 million. By 1932, as the drought took hold, sales were down to just $1.2 million.

made and hard decisions taken.* Russell himself was willing to make a start. He would cut expenditures, organize the collection of receivables, slash inventories and attempt to pay off the Commerce and other banks.

Despite his heavy commitment to Massey, Russell did not want the top management position. Instead he started looking around for a chief executive. He approached several people for advice, among them Vincent Massey in Washington. Vincent unhesitatingly recommended the young man whose name he had brought to the attention of the board in the past. In a letter to James Duncan in Paris, he told him of the advice he had given Russell: "I have recommended that you should be brought over from Europe and appointed Chief Executive Officer of the company."

Almost at the same time that he received the letter from Vincent Massey, Duncan received another from Russell summoning him to Toronto. He was thirty-eight years old. At the time he was co-manager with Cecil Milne of the French and German subsidiaries, and the European sales organization, then the only jewel in Massey's tarnished crown. Could it be that he was about to grasp the top job at Massey, with the power and authority that went with it? Duncan himself professed to think that he was still too young and inexperienced. But for a person as sure of himself and as confident in his own abilities as Duncan was, the prospect was exciting.

Once in Toronto, Russell told him that his name had been considered, but that he was in a key position in Europe, and the board of directors had felt that he should be left there and an outsider should be appointed to the job. What was needed, said Russell, was "a man experienced in manufacturing but one who has no ties of association or loyalty to the staff; a ruthless operator who will have no inhibitions about cutting down the organization regardless of long services, loyalty or past contributions. In other words, a surgeon and not a doctor". The job was to go to an American, the first to attain a top management position at Massey.

Not the figures but the people.

The new general manager, Bertram W. Burtsell, had no experience with farm machinery, but he possessed all those traits Russell had wanted in the new president. Burtsell was a tough

---

* The record of losses in the 1930s was enough to make any banker alarmed; the company lost $2.25 million in 1930, $4 million in 1931, $3.8 million in 1932, $3.3 million in 1933, $2.2 million in 1934, and $1.4 million in 1935.

operator with a reputation for ruthlessness. He had been production superintendent for the Packard Motor Car Company, a
company famous for its rough ways and its unorthodox labor
practices. Burtsell's appointment had been approved by the
board. A few days later he arrived in Toronto from his home-
town of Niagara Falls in New York state, a cheerful extrovert
who wore checked suits and delighted in describing himself as
someone who had been educated in the school of hard knocks.

Russell had asked Duncan to take the new general manager
with him to Europe and help him familiarize himself with
Massey's foreign operations. And so the two of them set out
together; Duncan, cultured and elitist, as fluent in French,
Spanish, and German as he was in English, and Burtsell, flashily
dressed, convivial, yet obstinately refusing to adapt to the
manners or customs of any country but his own. Surprisingly,
they got on well together. Burtsell was straightforward with
Duncan and Duncan admired the blunt decisiveness of his new
boss. Burtsell, he observed, got things done.

The first task that the new general manager undertook in
Paris was to fire Cecil Milne. He did so on the grounds that
only one man should be in charge, and that Duncan should
be that man. When he was shown profit and loss statements,
he waved them away. The important thing in running a business,
he counselled, was not the figures but the people, and the
important thing with people was to confront them face to
face and to see for yourself if they merited respect and were
worth employing. Like a whirlwind, Burtsell went through
Europe. In France, Germany, and England, he would visit
factories and sales offices and, after only a cursory inspection,
start criticizing the way things were being done or the
people who were doing them.

Canadian employees at the factory at Marquette les Lille
were treated to the Burtsell style. With no knowledge of the
personnel and little awareness of their respective jobs, he called
them into the manager's office one by one. Each was then given
a prepared speech that went: "Reach for your hat and coat my
friend, and get yourself and your family on a boat back to
Canada. You are finished with our company. Good luck to you
somewhere else. That is all. Thank you."

In due course, having completed his mission in Europe,
Burtsell returned to Canada to perform the same task. Not
unexpectedly, as the staff thinned out and jobs became scarcer,
the company's costs were cut back. Burtsell and Russell were
doing what they had promised. Still, little progress was made.
The recessionary conditions that had started in North America
were spreading around the world, and sales and profits showed
no sign of recovery.

Not all markets were affected with equal severity. But an important market which suddenly and dramatically collapsed was Argentina, and concern over the abrupt halt in sales there prompted head office to suggest that Duncan — while retaining his post in Europe — might be dispatched to Buenos Aires. Duncan had taken on a former Ford manager in Spain and France, Ken Hyslop, as his assistant general manager in Paris, and this enabled him to take up the Argentinian appointment. In the spring of 1932 he arrived in the Argentinian capital.

True to form, Duncan wasted no time. His ship docked in the morning and, on the afternoon of the same day, he was in the Massey-Harris office initiating salesmen and staff in their duties as he saw them, and attempting to raise morale. He refused to lower prices despite high inventories of unsold equipment. Instead he launched into a massive selling campaign, supported by field demonstrations and barbecue dinners known as *asado* to which all the farmers in the locality were invited. The technique worked. There was a revival in sales.

This success was accompanied by another that was less noticed at the time but of great importance in the long run. A company engineer, Tom Carroll, knew of a small Italian firm in the Argentine that had produced a revolutionary reaper-thresher. Instead of being drawn by horses, it was driven by an engine mounted on the chassis. Duncan and Carroll went to see the prototype for themselves. Then they borrowed the idea and developed it in Toronto. From there, Carroll returned in the following year with his version of a self-propelled combine. Tested in secret, it was put on the market around the world several years later. It was an instant success and ranked as the biggest breakthrough Massey had achieved in introducing new farm machinery. Competing companies jumped in, producing their own models. But Massey-Harris had a three-year lead in the market. So significant was this lead that for many years afterwards the major share of self-propelled combine sales in both Britain and the U.S. remained in the company's hands.

An ability to impress others.

In every post he had filled, James Duncan had done well. He had helped his father to safeguard Massey interests in Europe during the First World War; he had built up sales there in the postwar years; and, during his three-year stint in Argentina, he had achieved a substantial turnaround in business.

All this was to his credit. But in the mid-1930s, as in earlier years, the glamor attached to Duncan's name sprung less from his prowess as a salesman and his industriousness on the job than from his ability to impress others.

To the managers and directors of the firm, drawn as they were from the comparatively small-city environs of Toronto and its business establishment, Duncan epitomized in part their own aspirations. Well versed in the affairs of the world, he was also a man of personal charm and intelligence. At the age of forty-one he married a vivacious and talented Spanish dancer seventeen years his junior, Victoria Martinez Alonso, known as Trini*.

To the financial men and accountants running Massey, Duncan was almost a mythic figure and, as the Depression worsened, the repository of a great deal of hope for the future. Bertram Burtsell was recognized as a tough man, who was needed to do a ruthless job. But he was not a man to admire nor did he fire the imagination as Duncan could.

So it was that, in the winter of 1935, when the post of general sales manager became vacant, Duncan was contacted in Buenos Aires and urged to accept it. The position was second only to that of Burtsell, and it marked Duncan as his chosen successor. Twenty-five years after he first came to Toronto to work on the factory floor at Massey-Harris, Duncan prepared to return to Canada — a country he had only lived in for a scant four years.

After he arrived and familiarized himself with the head office and its staff, Duncan's first reaction was one of shock. In Europe and the Argentine, the Depression had struck hard and undermined confidence. But it had not produced the despair or sense of finality that gripped Massey's North American operations in the 1930s.

Although Duncan's first impressions were hastily formed, they were nonetheless uppermost in his mind when, on the second day after his arrival, he was invited to have lunch with Burtsell in his top floor suite at the Royal York Hotel. Before lunch the two of them motored along King Street in Burtsell's shiny eight-cylinder Packard, a new acquisition he wished to show off to Duncan.

Towards the close of the meal, Burtsell began to talk about his plans. He had a reason for wanting Duncan to come back to Toronto and that was to take over his position as general manager because his health was not good. But before he stood

---

* Duncan first met Trini on his sea voyage to Argentina; she was travelling with a Madrid company to play the winter season in Buenos Aires. A previous marriage to an Austrian girl in 1923 had ended in divorce.

down, the two of them had a mission to perform, and that was to implement a final retrenchment plan which had already been put to the board of directors.

Burtsell then proceeded to outline his plan of action. As the company had been unsuccessful in the U.S., it would cut its losses and cease to do business there. Burtsell asserted that the factories abroad were too costly to maintain, so he proposed a complete withdrawal from continental Europe, leaving just a sales office in England. The only place where Massey would continue to manufacture farm equipment would be Canada where it would have three branches, one in Ontario, one in the Atlantic provinces, and one in the west. Massey-Harris would, in effect, become a wholly Canadian company, although it would continue to export to a few other overseas countries where business was good, such as Argentina and South Africa.

Duncan listened to all of this with mounting disbelief. He did not interrupt, but he was appalled. Burtsell's plan contradicted everything he had stood for in the company; it made his two decades of building up effective sales organizations in Europe and South America utterly redundant. In particular, it flew in the face of all the hopes and dreams that two generations of the Duncan family, *pére et fils*, had nurtured in turn-of-the-century Paris. At the end of Burtsell's speech, he was asked for his opinion.

Duncan stated that in his opinion just about everything was wrong with the plan. He was confident that business would pick up in the U.S., and knew for a fact that Massey was in a strong competitive position in Europe. All these opportunities would be missed. Moreover, if manufacturing was ended in the U.S., the company would no longer have a line of tractors. Without tractors Massey could not maintain its share of business in Canada. Finally, seeing that Burtsell was becoming angry, Duncan told him bluntly that, rather than consent to the dismemberment of the company he and his father had been associated with for much of their working lives, he would quit.

Burtsell flew into a rage and swore at Duncan. He accused him of being ungrateful and of betraying his trust in him. He rose from the table, smashing a dessert plate on the floor. Then he stalked from the room, slamming the door behind him. Outside the hotel, he climbed into his gleaming Packard and drove home to Niagara Falls.

The next morning Duncan was called into the office of T.A. Russell who wanted to know why, a mere forty-eight hours after his arrival, he had quarrelled with Burtsell and refused to carry out his policies. In part, Burtsell's policies had been approved by the board of directors. In a panic over the continuing losses, which had now run for six consecutive years, the board had agreed that Massey's most problematic operations, those in the U.S., should be closed down. They were also in general agreement with Burtsell's view that all Massey operations should be trimmed down in keeping with its much reduced financial state.

Russell, however, had been none-too-happy with Burtsell's plans. As he listened to Duncan's argument that the proposed policy would destroy the company and prevent it from ever resurrecting itself, he was partially won over. As a compromise, he suggested that Duncan should go to Racine, Wisconsin, and look over and report on the U.S. operations. Meanwhile Burtsell did not press his case. Duncan's opposition so upset him that he stayed in Niagara Falls.

Burtsell may also have sensed that this could be a losing battle for him. While he was not a man to back down from a fight, he knew that ranged against him was the formidable hold that Duncan exercised over the company's directors and the favored status that he enjoyed.

On his return from Wisconsin, Duncan produced a report that blamed supply problems for many of the U.S. company's failures and suggested that, if the plants and branches were given the support and equipment they needed, it would be possible to cut down on existing losses and even operate profitably. Russell then asked him to look at the worldwide operations and prepare a budget for the following year. To make his point, Duncan had to show that without closing down any factories or branches, losses could be minimized. By his reckoning, this was possible. He estimated sales would increase and losses would be held to $250,000.

Since the company had lost $1.4 million in 1935 and since its U.S. operations had accumulated losses of $18 million and had never been profitable, Duncan's report seemed optimistic. Nonetheless, Russell considered it important enough to warrant calling a special board meeting to discuss it. In the absence of Burtsell, who continued to stay away from the office, he may also have felt duty-bound to give Duncan a fair hearing.

The board meeting when it came was chaired by Russell and attended by Burtsell. Without any formalities, Duncan entered the room and began to make his presentation, pleading with the directors for an opportunity to prove his case and arguing that to lose faith in the company now would be fatal. There were a dozen directors in the room, but the man he had to win over was Harry Gundy, who heard him out without saying a word. In the eight years since Gundy had committed himself to buying into Massey, the company had only been profitable twice.

By background, Gundy was a salesman and a masterful

one. He was used to winning over clients by appealing to their sense of vision, and by convincing them of the great opportunities that lay ahead. Duncan was by no means as persuasive and, judging by the company's record of the last few years, his case was a weak one. Nonetheless, he put it forward with his customary self-assurance, and Gundy was impressed. When it came to Burtsell's turn to speak, there was a touch of disdain in his voice. If the board followed the wrong-headed advice of Duncan, then the result would be bankruptcy for Massey-Harris. Moreover, he — Bertram Burtsell — would resign immediately.

After hearing the two men out, Gundy turned to Duncan. He agreed that if the company liquidated many of its holdings, as Burtsell proposed, then the losses would be very much greater. Duncan had said he could guarantee, with no closings, that losses would not exceed $250,000. Very well, said Gundy, if he had a promise that Duncan would resign the next year if his commitment was not kept and losses were over $250,000, he would give him the chance to prove his point.

It was a courageous decision, though both men must have felt they had little to lose. Gundy was so heavily involved financially in the company that it was easy to choose between a counsel of despair and one that held out some hope. For his part, Duncan would not have wanted to continue working for the truncated, solely Canadian company that Burtsell had recommended. Moreover, he was self-confident enough, and savvy enough, to have no doubts that he would succeed.

6

No Business like
War Business

*The war was to change the economic landscape
of the country entirely. It was to create within
a short space of time a newly industrialized
nation out of what previously had been a
country that earned its livelihood on the farm
and in the mines.*

For James Duncan, the attainment of the general manager's
position, the seat on the board of directors that followed, and
the day-to-day challenge of directing Massey-Harris from the
offices in King Street were the culmination of the great ambitions
of his youth. The problems did not overawe him. Nor did he
experience any qualms about the serious plight that the company
was in. To his associates he would talk about the task of leading
Massey out of the wilderness, plainly a tough responsibility.
But not for a moment did he doubt that it was his mission to
perform, and that he would make a personal success of it.

The problems did indeed run deep. Duncan could attempt
to solve them by gathering a new team of managers to run the
company. He could energetically raise morale, and talk
confidently about the future. And he could do his best to
concentrate the company's depleted resources in priority areas
that had been neglected, for example, bringing new model tractors and combines onto the market. But in reality, the company
Duncan commanded after six years of losses and depression was
a shadow of the prosperous global firm that had kept expanding
through the last years of the 1920s. Sales had collapsed as
completely as profits. In 1929 sales had almost touched $44 million.
By 1932, they were down to $9 million and, four years later,
had only recovered to $16 million.

Most telling of all for Massey was the virtual decimation of
its export markets. Among all the farm machinery companies,

it was Massey that had established itself internationally; its reputation had been built on its ability to cover the world. As the 1920s gave way to the 1930s, this process of internationalization came up against political and economic forces that compelled foreign countries to raise barriers to trade and obstacles to the importation of farm machinery.

Ripped apart by economic troubles that were soon to be transformed into political extremism, Germany banned farm machinery imports altogether. In 1934, Italy decided to conserve foreign exchange through a quota system. In the case of Canada, the quota that might have gone for farm equipment (since Canadian farmers did buy Italian farm machinery) went instead to purchase scarce supplies of nickel, copper, and steel. Denmark did not like the trade preference that Britain gave to Canadian bacon, so in retaliation the Danes set quotas on Canadian farm machinery so low that all trade virtually ceased. Greece prevented any foreign farm equipment from coming in whereupon its Balkan neighbors insisted on putting the barriers up too.

This tit for tat was not always hurtful; many of the markets were to small to make much difference. Massey-Harris continued to reach into France, and some other parts of Europe, from its manufacturing plant at Marquette les Lille. Trade with the countries of the British Empire was kept alive because of preferential agreements. Still, by 1934 Canadian exports of farm machinery had sunk to under $2 million from over $18 million in 1930. The fall in exports — combined with the precipitous drop in markets at home — was damaging to Massey, and therefore a major worry for Duncan.

The woes the company faced in the U.S. and Canada added to its troubles abroad. The effects of the Depression on Massey's North American operations had been calamitous, and management had been left with no alternative but to prune costs rapidly or suffer even greater losses and eventual insolvency. It was the disastrous position in the U.S. that had persuaded the board to agree to Burtsell's proposition that the Batavia plant should be demolished to save taxes; that the machinery should be shipped back to Toronto; and Massey's operations in the U.S. liquidated. Duncan had gambled by opposing this plan. And he had won. But winning in the board room in Toronto was not going to boost Massey's product line or give it a strong marketing organization in the U.S.

When Duncan had triumphed and Burtsell had stalked off, almost his first task was to order a halt to the removal of machinery from Batavia. In itself the action was symbolic. It showed Massey employees in the U.S. that the company was not running away and that it would stay put. But it was almost

a last-ditch effort. The plant at Batavia was rundown and in
disrepair; its products were noted chiefly for their poor quality.
Neglect by head office had reached such proportions that the
company in the U.S. had no sales manager. The plant's director
had to double as the chief of sales, and the duty was an onerous
one. The task of selling Massey products in the U.S. was not
easy. The company had lagged behind its competitors in the
change to tractor power and in the 1930s this meant that
much of the business in the U.S. corn belt was lost to others.*

The debacle in the U.S. was a challenge to Duncan. In actual
fact, however, he was relatively unaware of just how serious the
American situation was. His decision that Massey must stay in
the U.S. at all costs had been taken on a hunch that markets there
would improve. While in the Argentine he had noticed a small
market developing for surplus tractors from the U.S. and he took
this as a sign that U.S. farmers were beginning to buy new equipment. But plainly a much more substantial recovery was needed
to revive Massey-Harris. Duncan, with his background in Europe
and South America, had no firsthand experience of the vagaries
of the U.S. market or the fierce competition that Massey faced.

To go with the troubles over manufacturing, there were
financial troubles. Aware of Massey's continual losses, and the
hasty reversal of its decision to cut and run, the company's
New York banks were reluctant to extend more credit. Nor were
there other sources to turn to. The lamentable record of Massey
over the previous six years had created a shortage of working
capital and made it vitually impossible to raise new funds.

This was the impasse that Duncan found himself in. The
line of farm equipment that Massey was selling was out-of-date
and in need of refurbishing. But this could only be done if
investments were made in the company, generating more money
to spend on developing new products for both the Canadian
and U.S. markets. Where was this money to come from?

In early 1936, one solution seemed obvious to Massey and
the other farm equipment makers. As they had been hit by a
jump in steel prices, they reasoned it would be feasible to
take advantage of a slight increase in the demand for their
products and raise their prices. For nearly all the machinery
makers, costs were above what they could afford to charge.
Farmers had been so impoverished by the Depression that to

---

* In the 1920s and 1930s, there were numerous improvements in tractor design as U.S.
manufacturers and U.S. farmers led the change from animal to tractor power. International Harvester had successfully introduced row-crop tractors, with high clearance
and front wheels positioned close together, as early as 1924, and a line of general-
purpose tractors which could be used in some row-crop field-followed. Massey-Harris
did not emulate its rival and, consequently, could not sell a full line of row-crop
implements. In addition, it did not move into high horse-power tractor or crawler
tractors.

make a sale, any sale, the firms were resigned to taking a loss. Economically, therefore, there was plenty of justification for the price rise, but, politically, such a move was fraught with peril.

The old complaints about the exploitation of farmers by the implement makers had not been stilled. To this had been added the grievances of farmers in western Canada, the victims of five years of falling crop prices and poor harvests. For the farm companies to raise prices now was bound to raise an uproar. And so it proved.

The Liberals under Mackenzie King had just swept to power, winning 171 seats in Parliament. Only 39 Tories from R.B. Bennett's governing party survived, and their leader himself had departed.* Mackenzie King wanted to rally western Canada behind the Liberals to give the party a truly decisive victory. So he rapidly moved to support the western farmers' cause and criticized the farm implements industry. First, tariffs on U.S. imports were stripped down, with duties on tractors being abolished altogether. Second, an inquiry was launched into the farm implement industry in Canada and its pricing policies.†

After hearing testimony over many months the two House of Commons committees investigating the industry concluded that the 1936 price rise had not been justified; that the losses suffered by the companies in the 1930s were due to farm incomes being depressed and that prices for farm machinery during all these years had been too high; and that the government should take action against the companies. Having drifted through its work at a leisurely pace, the second Commons committee had in fact outlasted the issue. Passions had cooled, and no action was taken to put its recommendations into effect.

James Duncan had been among the witnesses called before the committee. Surrounded by a group of aides who plied him with figures, refusing to be badgered or hurried in giving his testimony, he put on a skilled performance in defending Massey. So impressive was his defence that an audience of MPs gathered in the hearing room; they were there, in the words of one of their number, Bill Fraser, the MP for Trenton, Ontario, to "get a kick out of hearing two-bit Western MPs trying to take on a one hundred thousand dollar executive." Duncan had made his mark in Ottawa.

---

* Bennett had taken the title of Viscount of Mickleham, Calgary, and Hopewell and gone off to live in England as the neighbor of another lordly ex-Canadian, newspaper magnate Lord Beaverbrook.

† Aside from Massey, the two largest companies in Canada were the Canadian subsidiary of International Harvester and the Cockshutt Plow Company.

For all the fine words, the price inquiry and the lowering of tariffs had been a defeat for Massey rather than a victory. The big American companies were able to sell more in Canada as a result of the tariff changes. And the price inquiry, inconclusive though it was, had warned the industry off future increases. In 1937, a drought held back business in the west once again and caused more distress. But gradually, imperceptibly at first, a recovery did begin to take shape. For Massey it was not enough to put business back on a firm footing. The company had to be content with making limited progress. By 1937, after eight years of losses, only a start had been made — Massey only managed to make a profit of $1 million.

To actually earn any money at all at a time when the company continued to be short of capital and unable to expand, was a breakthrough. What was beginning to have significance — and this was something new for Massey after a long period of time — was a major improvement of its products. Pushed along by Duncan, the company was at last beginning to come up with competitive new models. Chrysler automobile engines were incorporated into a new line of tractors. The Massey-Harris Clipper combine was developed. Manufactured in Batavia, the successful combine managed to turn a loss-making plant into a profitable one. Most important of all, Massey launched the self-propelled combine in 1941. Despite the onset of the war, it provided a major boost to sales in the U.S. and Canada — and marked the company as an innovator in international markets, a reputation it had not enjoyed since the days of Hart Massey.

Still, the gains and the breakthroughs were limited. In taking up the presidency, Duncan had insisted that Massey remain an international company and not become purely Canadian. True to this spirit, and to its origins, the company kept trying to export and to sell abroad even as the situation in Europe deteriorated and war became imminent.

The Second World War was to change Massey and to transform the company into a far larger manufacturing enterprise. But, on the eve of war, four years after Duncan had emerged as general manager, the results and achievements of his term had been mixed. He had been held back by weak finances and capital shortages. He had faced uncertain markets and had done his best to develop new products. He had kept the company intact and refused to draw back from the U.S. despite great difficulties.

All this was on the plus side. But there were negative factors at work and Duncan, for all his drive and personal magnetism, had been unable to reverse them. Slimmed down by the Depression, Massey had lost markets in Canada and abroad while in the all-important U.S. market it ranked no higher than seventh. Political and economic events had conspired against a company that had always striven to be international in character. Lack of managerial foresight and knowledge in the 1920s and 1930s had put it behind in the tractor market. All these were not hurdles that could be readily overcome. Nor could different managers have changed things very much. As the war closed around Massey, it was apparent that the firm had managed to survive the Depression, albeit in a weakened form. What it had lost was the momentum of the early days. Massey was farm from being a prosperous company, and its years of exuberant expansion seemed to be a thing of the past.

Only four months after Britain and Canada declared war Massey was producing anti-aircraft shells.

At the outbreak of the war, Canada was still a relatively small nation, with a population of eleven million, and it was a nation that was suffering economically. A full decade after the beginning of the Depression, there were 600,000 people out of work. The search for jobs and for employment stretched across the country, with much of the worst hardship being felt on the Prairies. In the big cities of the east, manufacturing industries were showing the first signs of growth and expansion. But the total number of Canadians employed in manufacturing was only 650,000, scarcely more than the number of jobless workers.

The Second World War was to change the economic landscape of the country entirely. War meant jobs for the jobless. War also helped build up the economy. As the needs of Britain and later of the Allied Forces turned to material as well as men, a newly industrialized nation was created within a short space of time out of a country that had earned its livelihood on the farm and in the mines.*

The initiator of this was Mackenzie King. Fearful of the divisive effects conscription might have on Canadians, he preferred to put the stress on one of the side-effects of war; the full-employment economy that it made possible. But while King gave his support to the war effort and the links that were forged with wartime Britain, the actual architect of the industrialization policies was the most powerful member of his cabinet, C.D. Howe.

* The war industries by themselves were to remould Canada's economy, creating jobs for one million men and women, doubling the gross national product (GNP), and boosting Canada into the artifical position of having (by the war's end) the free world's third largest economy.

Born in Waltham, Massachusetts, Howe had graduated in engineering from the Massachusetts Institute of Technology, become a professor at Dalhousie University, then moved to northern Ontario as chief engineer with the Board of Grain Commissioners before starting his own engineering firm which built grain elevators around the world.

Howe was tough and pragmatic and a believer in the free enterprise system. He was also a man who cultivated the rich and powerful, and who had no difficulty in becoming assimilated into the upper strata of Canadian society. (It had been Vincent Massey, then president of the National Liberal Federation, who — over dinner at the Chateau Laurier Hotel in Ottawa in 1933 — had persuaded him to enter politics). Howe liked to use his influence to get things done. And he had an enormous appetite for accomplishing things. As the minister in charge of war production, he cut through the barriers that had often stopped governments co-operating with business and established an autocractic command over the nation's economy. Howe wanted capable men to run the war effort. He would parcel off to them the task of managing Crown corporations and whole sectors of the economy. It was an efficient way of doing things, and to Howe's mind efficiency was everything.

When the war started, Howe had been an MP for just four years. Elected as the member for Port Arthur in the Liberal sweep of 1935, he had immediately gone into the cabinet at his own insistence.* Once in Ottawa, he had presided over the establishment of two national institutions, CBC and Trans-Canada Air Lines, later Air Canada. He had also, in prewar days, forged close relations with the business elite of Toronto and Montreal. Prominent among his new friends and allies was James Duncan.

Howe's links with Massey-Harris were to become important, both for the company and for the national war effort. In the fall of 1939, with war already declared and the first British delegations arriving in Ottawa to seek Canadian aid, there were few individuals or companies either prepared or equipped for hostilities and for the necessary enormous conversion to wartime organization. In Britain the national psyche had not yet adjusted to the reality of being at war, nor was there any appreciation of the demands that would be put on the country. In Canada, the mood was even more relaxed.

This was not, however, the case with James Duncan.

* Doing deals to enter the Mackenzie King cabinet was quite a common method of advancement for men of influence, and those on King's list were often prominent businessmen whose reputation would help enhance the prime minister's own political standing; not just C.D. Howe but also Vincent Massey and James Duncan were asked to join the cabinet. The understanding was that they would join the cabinet after being elected in a safe Liberal riding.

Attuned to events in Europe, he had before the outbreak of war dispatched the company's chief engineer, Guy Bevan, to England to make known the manufacturing capability Massey could offer the British. Steps had been taken to order machine tools, hire skilled workers, and put in place a small staff to handle war orders. Four days after Canada declared war, Duncan was in C.D. Howe's office in Ottawa briefing him on Massey's factories and the type of defence work that could be done.

In many ways, Duncan and Howe were men of a similar mould. Both hated inaction and indecisiveness. Both were self-motivated and ambitious, hard-driving and egocentric. Such similarities might, in other circumstances, have forced the two of them apart. But they had a shared background as outsiders who had come to Canada and achieved prominence on their own. And they had a common bond in their belief in the country and in its potential for greatness, something which both of them felt would be realized as a result of the war and the contributions the country could make.

The two of them became friends. Duncan proffered advice to C.D., as Howe was known, and, in the fullness of time, he joined the exclusive ranks of the dollar-a-year men whom Howe recruited to oversee the war effort. The select group were to preside over the wartime industrialization of the country, and thereby attain an ownership position later when Canada entered a period of postwar prosperity. They were to attain for themselves invincible positions at the head of its major corporations.

But all that was in the future and not something that much concerned either Howe or Duncan. Both men were organizers and managers, and did not greatly desire to be owners. What they both needed and wanted were positions of power. And in wartime Canada, power on a scale that had never been known before was being vested in the hybrid team of administrator-businessman-bureaucrat advisers who were being given free rein in Ottawa.

Under Duncan, Massey-Harris was ready and available for war work. And since manufacturing capacity in Canada was limited, Massey could expect substantial orders. In January, 1940, only four months after Britain and Canada had declared war on Germany, the Toronto plant was producing 40 mm anti-aircraft shells. These were the shells that would defend London and help win the Battle of Britain. Other orders for shells came in following the conversion of the Woodstock plant. Both plants together began manufacturing bodies for truck and personnel carriers. Wings for the Mosquito fighter bomber and the Avro-Anson twin-engine trainer were made at Weston, and links for tank treads at Brantford. A new building was constructed at Brantford in 1942 to manufacture the mounts for naval guns needed to equip corvettes, frigates, and merchantmen for the war in the Atlantic. Before the end of the war, the Canadian factories had added 60-pounder shells, bombs, cargo bodies, radio trailers, and air-field tractors to their array of war material.

These orders, and the expansion they brought to the company's activities, arose out of the partnership that had suddenly sprung up between government in the person of C.D. Howe, and the business community. A job had to be done, and it was the free-enterprisers who were called on to get it done. For Massey-Harris, relations with the government in Ottawa had always been a subject of vexation and hard feelings. There had been the continual debates over tariffs and prices, and the allegations that the farm machinery makers were exploiting the poor farmers of the West.

Now C.D. would be on the phone to Duncan urging him to come to Ottawa to undertake a project for him.

Once there, Duncan would learn that his friend wanted him to draw up a plan to cover the allocation of raw materials to the farm machinery industry. The conflict of interest involved in the assignment was astounding. Duncan, as the head of the largest Canadian-owned farm implement maker, was to advise the government on how much in the way of scarce raw materials should be made available to the industry, and to his company. Duncan went ahead and drew up the plan (which was in fact quite restrictive), and C.D. Howe on behalf of the government and people of Canada went ahead and accepted it.

All through the war Massey-Harris continued to make farm machinery. Production had to be cut back because raw materials were short, but an exception was made in the case of exports to Britain — which was in critical need of food — where shipments of combines and tractors were increased. Business was booming, despite the fact that Ottawa levied an excess profits tax which kept earnings from rising. Altogether $3.5 billion was invested in new manufacturing capacity during the war, with much of it coming from the government in tax credits and allowances. Massey's sales climbed steadily every year: $21 million in 1939, $58 million by 1942, and $116 million by 1945. Of this sum, more than half was accounted for in war production (the peak year was 1943 when war sales accounted for $73 million out of total sales of $91 million).

If the war business was big in Canada and getting bigger, it was potentially enormous in the U.S. Because of the experiences of the parent company in Canada, the U.S. operation had a head start on its competitors in organizing itself for wartime production. Even before Washington had entered the war, the

Racine factory had received an order from the government to manufacture links for tank treads. And at the time of Pearl Harbor, two officers from the U.S. Army Ordnance department were in Wisconsin to look at the possibilities for manufacturing tanks.

The chief of Massey in the U.S. was Ken Hyslop, the former Ford manager whom Duncan had entrusted with the top job in Europe when he had gone to Argentina. Hyslop had joined Ford as manager for Spain and subsequently for France in the 1920s. Like all Ford managers, he had been told that before joining the company he had to sign a letter of resignation. One morning arriving at his office in Barcelona, he found a new man sitting at his desk. Hyslop was told that head office in Detroit had just accepted his resignation. He had in fact been a successful manager in Spain and was later told that it had all been a mistake. He was then rehired as Ford manager in France.

Mindful of Ford's erratic hiring and firing practices, Hyslop had later accepted the security of a job at Massey offered to him by Duncan. Hyslop was a larger-than-life character. Relaxed and amiable, he spent his leisure time tracking big game in Africa, hunting wild turkeys in Mexico, deep sea fishing in the Caribbean, and playing golf superbly; and he continued to do all these things until he was well into his eighties. Hyslop was second-in-command to Duncan and became president of the firm in the U.S.

On January 6, 1942, Hyslop sat down with senior officers from the U.S. Army Ordnance department and listened, calmly and silently, to the biggest and best deal that had ever come Massey's way in the U.S. Washington wanted to sign a contract that would involve managing the production of 1,200 tanks a year.

After the meeting, Hyslop tallied the return Massey could expect if it went ahead and signed the contract. Then he contacted Duncan. "The whole thing may seem too good to you to be true, just as it was to me," he wrote. A few days later a letter of intent was agreed on and, shortly afterwards, Massey-Harris bought the vacant Nash-Kelvinator plant in Racine to enlarge its factory space. The deal called for Washington to put up working capital and pay for machinery and machine tools.

Massey did not make full use of these grants. It was decided that it would be more economical, and quicker, to subcontract much of the component work. By June of 1942, three months ahead of schedule, the first M-5 light tank rolled off the production line.

The tank contract in the U.S. was worth $133 million during the war years, and accounted for about 60 percent of the contracts, and the expansion that it made possible, the tank deal benefited the company in two important ways.

First, it created an expertise in subcontracting that had never existed before; literally hundreds of suppliers worked under Massey's purchasing and subcontracting managers.* The second benefit was the recognition that Massey received in the U.S. Its expertise was commended by the U.S. government which called the Racine operation "an example of subcontracting on a prodigious scale that was one of the most outstanding in the U.S." It received the Army and Navy E production award. The struggling farm machinery company became known to a wide segment of U.S. industry and the public. By the end of the war, Massey's tank production had given the company a bigger profile than its farm implements, and one on which it would be able to capitalize on in the postwar years.

Acting deputy minister of defence for air.

In the spring of 1940 Duncan was on a business trip in California when he received an urgent call from the Canadian minister of defence, Norman Rogers. He was speaking on behalf of the prime minister, said Rogers, in asking Duncan to come to Ottawa and help organize the British Commonwealth Air Training Plan.

Mackenzie King was under fire. The war in Europe was being stepped up. And the press were beginning to ask questions about why the air training plan, one of the largest partnerships with the British, was not functioning better. Training centres were slow in being opened up. And there was a desperate shortage of trained pilots and observers for the air force.

Duncan at first turned the proposal down. He was preoccupied with the war contracts that Massey was getting — and hoping to get — in Canada and the U.S., and concerned about the health of Massey's president, T.A. Russell. Apart from this, he was unfamiliar with the air training plan, and with what the demands of the job would be. Rogers said he was disappointed and rang off. As soon as the conversation was finished, Duncan's wife, Trini, persuaded him to reconsider. Rogers was contacted again. The following day, Duncan was in Rogers' Ottawa office discussing the air training plan and his

---

* The complexity of the purchasing and subcontracting work can be seen from the fact that each shipment of 100 tanks had to be supported by spares which consisted of 190,000 different pieces shipped out in 3,657 different boxes.

appointment as acting deputy minister of defence for air. In on the discussions were C.D. Howe, who had suggested his friend for the job, and Prime Minister Mackenzie King.

Duncan could only take the job if the board of directors at Massey would agree. When he found out that the president, Russell, had been taken to hospital after suffering a second heart attack, Duncan was loath to commit himself and wanted to withdraw. However, other board members — on the urging of Howe — presented a united front. Duncan was told that he should accept the post in the public interest.

The move to Ottawa put Duncan among the dozen or so top administrators of the war effort and dollar-a-year men. Recruited from the ranks of the nation's entrepreneurs, "C.D.'s boys" billeted themselves in the Chateau Laurier Hotel, took their lunch at the Rideau Club or more often at their desk, and rejoiced in the camaraderie of working sixteen-hour days and wielding power on an extraordinary scale.*

Only a few weeks after Duncan was appointed, Norman Rogers was killed in a plane crash on his way to Toronto. For a time there was no replacement for him and then Colonel Layton Ralston was moved from his position as minister of finance to become the new minister of defence. A few weeks later, Duncan was handed a confidential memo from London predicting the fall of France. The situation in Europe was fast becoming critical. Britain now stood alone as the Luftwaffe mounted massive air raids. Nightly, 1,000 planes bombed London, Liverpool, and Coventry.

These events stirred the imagination and the sympathy of the Canadian public and led to calls for greater help. What was the point of long-term air training plans now? Would it not be better to offer immediate assistance, and send Canadian and Commonwealth airmen into action? The public mood swayed Mackenzie King; he ordered reports to be sent to the air ministry in London on alternative plans and on hurrying the training programs along. The British, however, refused to be panicked. They asked for more fighter and bomber aircraft. These were sent over immediately, though it meant leaving Canada's own coastline poorly defended. At the same time the British told King that the best help Canada could give would be to press ahead with a training plan that would supply tens of thousands of airmen from 1941 onwards.

---

* Among the outsiders in C.D.'s team were H.R. MacMillan of the H.R. MacMillan Lumber Company of British Columbia, responsible for timber and shipbuilding in the ministry of munitions and supply, and an executive committee that consisted of prominent businessmen, lawyers, and accountants: R.A.C. Henry, Gordon Scott, E.P. Taylor, W.A. Harrison, Henry Borden, Colonel Eric Phillips, and E.K. Shiels, the only civil servant in the group.

To support the program, C.D. Howe and Duncan went to
Washington to buy the training aircraft and instruments that
Britain could no longer provide. They also committed greater
funds, and thus brought forward all their plans. The new
airfields that were to be completed during 1940 and 1941 would
now have to be finished before the end of the year, together
with the training centres. The number of airmen to be trained
was substantially increased.

The work that Duncan was doing gave him a political
reputation as much as an administrative one. And it brought to
the fore the talent that he had always possessed for impressing
others, and for winning laureates from people of influence and
power. Less than three months after he had first come to
Ottawa, the proposal was floated that he should become
minister of national defence for air. Shortly after this he had a
chat with Mackenzie King who offered him a seat in the
cabinet.

Duncan demurred. He had never thought of being a
politician. Mackenzie King reassured him that there was
nothing to it. He personally would choose a safe Liberal seat
for him, and he suggested either Kingston or Waterloo or, since
Duncan was bilingual, a riding in Quebec. "All you have to
do," said King, "is make a couple of speeches. You will be
elected without any trouble and immediately be brought into
the cabinet."

Duncan pondered his decision for several days. To go into
politics would mean having to leave Massey-Harris. Fourteen
years earlier, Vincent Massey had offered him a diplomatic
position in Washington. That offer had been less tempting, and
easier to turn down. This one was much harder. He went to
Toronto and discussed it with the board. And he talked to his
friends and colleagues in Ottawa. A political career would have
put him in the public spotlight, a not unattractive proposition
for a man with Duncan's self-esteem and sense of his own
worth. But would it give him greater power, and greater
security in the use of power? Or was it just a matter of prestige?

The decisive advice for Duncan seems to have been that
passed on by his political friends. They warned him that
Mackenzie King's judgment of him was based on his success with
the air training plan, and his status as a Toronto industrialist.
As of now, he was King's blue-eyed boy. But once in the
cabinet, he would quickly become less admired. "You must
realize," counselled one of his colleagues, "that if anything
ever goes wrong in the months or years to come, the chief
would not hesitate one moment to throw you into the discard."
The advice seems to have made up Duncan's mind. He would
not hazard his life long career at Massey, and the chance it gave

him to rule over a global business empire, for the uncertainties of Canadian politics. He went to see King and told him that he had decided against the proposal.

Later, he must have felt his decision was fully justified. Colonel Ralston, his boss as minister of defence and a man who had served under King with distinction, had been one of the friends who had warned him of the dangers of political life. When the armed forces were in urgent need of reinforcements, Ralston urged that in the national interest conscription should be introduced. This was an issue that would have split the Liberal party. King prevaricated, went behind Ralston's back to persuade General Andrew McNaughton to take the defence post and support the concept of an all-volunteer army, then unceremoniously fired Ralston. It was not the kind of fate James Duncan wanted for himself.

Having turned down a future in politics, Duncan felt less reason to be in Ottawa. The air training plan was proceeding on schedule. Besides, there were many matters in the Massey head office on King Street that required his attention. He told King that he wished to retire from his position with the air force at the end of January, 1941.

After nine months in Ottawa, Duncan returned to Toronto. T.A. Russell, who had presided over the company in the 1930s, and promoted Duncan's ideas over Burtsell's in the crisis year of 1935, had died the previous month. The board of directors unanimously voted that Duncan should succeed him as president.

C.D. Howe's boys.

The attainment of the presidency was a natural progression from the job of general manager. With it as well came an extra piece of authority which was to be crucial in Massey's future; in the absence of a chairman of the board (a position that Duncan was to claim for himself a few years later) the sole discretion for recommending new members to the board of directors was to fall to the president — Duncan.

In the years 1941 and 1942 the ranks of the board of directors were depleted by the loss of western representative and Winnipeg lawyer, George Allen, investment banker E.R. Wood, and George McLaughlin, the brother of motor car pioneer Colonel Sam McLaughlin. In addition, a reorganization plan called for the number of directors to be increased from ten to twelve. Duncan was looking around for suitable candidates to be outside directors and take an interest in the affairs of the company. Having so recently been immersed in Ottawa, Duncan's thoughts naturally turned to the dollar-a-year men who had put their talents at the disposal of C.D. Howe.

One of C.D.'s boys, E.P. Taylor, was a neighbor of his. During Duncan's time in Ottawa, the two men had often discussed the future of Massey-Harris and Duncan's plans for the firm in the postwar period. E.P. Taylor was a rising star in the business community and president of Canadian Breweries, and he seemed to Duncan to be an ideal choice. So during a trip to Washington, where Taylor was doing his war work as the chief executive officer of the British Supply Council, Duncan invited him onto the board.

For Duncan, it was a fateful decision and one that he was to bitterly regret later. For the gregarious Eddie Taylor, the offer of a Massey directorship was an upward step and an opportunity to become better established and wealthier. Such opportunities were not to be missed. Taylor had the finely tuned instincts of a born speculator. If he was to sit on the board of Massey-Harris, his function would not be to offer sage advice to an all-powerful president and general manager, or to massage Duncan's ego: Eddie Taylor would be in it for himself.

In prewar days, Taylor had made his reputation by gathering together a group of money-losing Ontario breweries, consolidating them into Canadian Breweries Limited, and managing to get himself a marginally profitable company. Starting with the Kuntz and Brading breweries in 1929, he had merged seventeen brewing companies, financing his purchases with share swaps and, at a critical time, with help from British interests. In all, Taylor had reduced the number of beer brands from 200 to 9. By cutting and slashing he had jockeyed his Canadian Breweries into a position where, by 1941, it was sharing the Ontario market with John Labatt Limited of London, Ontario, and making a slim profit of $81,000 on sales of $12.7 million. Taylor had carried out his mission by doing one deal after another. He was not an experienced businessman, and he was certainly not a manager or administrator, but he had a great entrepreneurial flair and was hyperactive when it came to doing business deals.

He was also known to both Harry Gundy, the most influential director on the Massey board, and to C.D. Howe. At the time when the Liberals were considering a national airline, Taylor and Gundy had offered to start one using Douglas Aircraft's DC-1, later to be transformed into the durable DC-3. Howe might have gone along with the idea. But his Liberal colleagues decided a national airline serving the country from coast to coast had to be government-owned. The Liberals then

96   proceeded to pass legislation approving Trans-Canada Air Lines. After the airline venture, Howe knew about Taylor even though the two had never met.

The eventual meeting between Howe and Taylor took place at Ottawa's Rideau Club. It says something about the makeshift nature of wartime planning and social contacts that a chance casual meeting could vault Taylor into one of the most influential posts in the Allied war effort. The meeting took place in April 1940, exactly the same month as Duncan was being telephoned in California and prevailed upon to take his post on air defence work.

Taylor was in the Rideau Club having a drink with his father, Colonel Plunket Taylor.* Across from them was a group that included C.D. Howe; Montreal accountant Gordon Scott; and an old friend of Taylor's, Henry Borden, a Toronto lawyer. The nephew of a former prime minister, Borden had acted for Taylor on several brewery acquisitions. Now a dollar-a-year man, he was discussing with C.D. Howe the need to attract talented executives to run their fast-growing war machine.

Borden came over to Taylor and said that Howe would like to meet him. After they were introduced, C.D. came straight to the point. "We could use your help," he told Taylor. A few weeks later, E.P. Taylor signed on as perhaps the capital's youngest dollar-a-year man, moved into a room at the Chateau Laurier, and started work at the building on Wellington Street that housed the top men of the ministry of munitions and supply. Along with Scott and Borden, he was a member of Howe's executive committee, and was given responsibility for procurement of all munitions and armaments for the Canadian armed forces, and for the British. Initially, the British side of his responsibilities did not have much significance. The government in London did not want to place orders in Canada; it wanted to manufacture its own equipment and munitions and, if possible, supply the Canadians too. However, as the situation went from bad to worse in Europe and Britain found itself isolated and its factories bombed, Canadian production began to become more important.

Increasingly, the British war effort had to be kept going through supplies that could only be obtained in Canada and

* Taylor's father, Plunket Bourchier Taylor, had been born in Kingston in 1863. His mother Lucy's family had included some prominent English admirals and generals (hence the grandness of his names), and the young Plunket had joined the militia and helped put down the Riel Rebellion before marrying Florence Magee, the daughter of the founder and president of the Bank of Ottawa, Charles Magee. He then went to work in his father-in-law's bank. In the First World War, Plunket Taylor became a colonel in charge of the Canadian Pay Corps in Europe. Later the family bank was bought out by the Royal Bank but other Magee family interests were kept up; these included Brading Breweries which grandfather Magee had acquired. Plunket Taylor had been a director and president of the company and young Eddie Taylor was to use it to found his brewing empire.

the U.S. Taylor became the agent for negotiating with Canadian 97
manufacturers, cutting through red tape and, often on the basis
of a hastily scrawled letter or memorandum, committing himself
and the government to multimillion dollar orders for the
Canadian armed forces and for the British.

Because of its importance, the work could only add to
Taylor's stature. Within a year, his ability to do deals and initiate
new ventures was as well recognized in London as it was in
Ottawa. The signing of the Lend/Lease Act under which the
U.S. would supply Canada with the war materials it was
shipping to Britain created another job for him. Henry Borden
had established a Crown corporation, War Supplies Limited, to
negotiate with Washington for war supplies and manufacturing.
Howe called Taylor into his office and told him: "Eddie, we've
formed a company called War Supplies Limited. You're the
president. Go down to Washington and sell it." Taylor set up
his office in the Willard Hotel and began to do an energetic
sales job in the U.S. Later he was to extend his work with
the British, becoming chief executive officer for the British Supply
Council in North America.

These activites and other like them were exciting and time-
consuming. In his background, E.P. Taylor was not untypical
of all the dollar-a-year men Howe recruited for the war effort.
While some were civil servants, lawyers, and accountants rather
than entrepreneurs, they had all been chosen for their capabilities.
C.D. Howe was to say of them all that he did not give orders
to his boys, he only gave them responsibilities.

The network of contacts established among the dollar-a-
year men was to be a lasting one, and would leave its mark on
the country and on the participants. When E.P. Taylor met
James Duncan, the latter, along with H.R. MacMillan of the
H.R. MacMillan Lumber Company, was the best known
business executive working in Ottawa.

In Ottawa Taylor was also able to renew an old friendship
with Colonel Eric Phillips. Just as Taylor had been invited into
the inner circle because of his acquaintance with Henry Borden,
so he passed Phillips' name along to Howe. The opportunity
came when Howe was looking for someone to assume responsibility for the procurement of guns for the navy. Taylor
recommended his friend on the grounds that Phillips was a good
businessman and knew all about boats. Phillips had one yacht
moored in Georgian Bay and another in the Bahamas.

Phillips did indeed have a passion for boats, so much so
that when he could not get a company to repair his favorite
motorboat satisfactorily, he acquired the company and made
sure the work was done. Tough and aggressive, Phillips was
a far better administrator than Taylor would ever be. Eight
years older than Taylor, Eric Phillips had been born in Toronto

and educated at Upper Canada College and the University of Toronto where he had gained his bachelor of science degree before the outbreak of the First World War. Phillips came from a moderately prosperous family that had a small glass business. After travelling in Europe, he had joined the British Army, first in the Leinster Regiment and then the Royal Warwickshire Regiment. His military career had been a distinguished one; a colonel by the end of the war, he had been awarded the Military Cross and the DSO, had twice been mentioned in dispatches for conspicuous bravery, and had been wounded on the western front. For a brief period after the war, his services were loaned by the British to the French for administrative work in Poland. He had retired from the army and returned to Canada in 1920.

In the years between the two wars, Phillips had built up the family business, Duplate Canada Limited, taking it in new and different directions as a supplier of glass to the booming auto industry. The colonel, as he liked to be called, had also forged close links with the business establishment of the time. His first wife had been a daughter of Colonel Sam McLaughlin, who had sold his motor car company to General Motors, and whose brother was on the Massey board. Once in Ottawa, Phillips made himself familiar with the men of power, befriending politicians and civil servants. He also developed a power base of his own by taking charge of a Crown corporation, Research Enterprises Limited, which had been formed to plan wartime industrial development.

Taylor and Phillips became close associates though they could frequently be seen arguing, sometimes good-naturedly and sometimes not. Both men instinctively recognized the special qualities that each possessed. Taylor could not match the colonel in his organizing skills or his ability to motivate people. Nor did he have the same shrewd judgment. Phillips on the other hand saw in Taylor a great capacity and energy to do new things, to make deals, to set himself goals and pursue them ruthlessly. If there was one overriding reservation Phillips had about his younger friend, it concerned Taylor's sense of adventure and his great schemes for the future. Eddie had to be brought down to earth. Sometimes his schemes were altogether too fanciful for the colonel's taste.

During the early war years, Taylor and Phillips did not have any formal business association. Nor did they discuss becoming partners in any serious way. Nonetheless, the mutual respect they developed provided the framework for what was to follow — the creation of the country's most famous investment firm, Argus Corporation. Given the informal alliance that existed between the two men, it was not surprising that when his confrere should follow him.

Taylor took his seat on the board in March, 1942. Phillips was on the board by December of the same year. Both had been welcomed by Duncan who genuinely wanted to upgrade the quality of board members, and felt the two men were suitable candidates. Their arrival did not change the way that Massey-Harris was being administered and run for the time being. Duncan had gathered all power to himself by taking the post of chairman of the board as well as president and general manager. But power was to shift in a way that had not happened before in the company's history. With the formation of Argus Corporation in the first months after the war, the opportunity to win control of the company, (an opportunity that had existed since Vincent Massey's 1927 sellout) was about to be seized. For Taylor and Phillips it was too obvious an opportunity to be missed.

Argus takes over.

Taylor and Phillips had begun to buy Massey stock as early as 1943. At the time, they had only a hazy idea of how this investment, and others they were making, would fit together, but they were prepared to form a business partnership and had taken a joint shareholding in two companies that Taylor had spun off, Canadian Investment Company Limited (Caninvesco) and Invesco Limited. The strategy that shaped their investments was a simple and opportunistic one. The war was coming to an end and share prices were generally depressed as a result of uncertainty. In the case of Massey, few investors were prepared to take a chance on the company's success in adapting to postwar markets following the loss of its rich war contracts. Taylor and Phillips, as directors, had no doubts about the company or its future. They were both willing to speculate that markets would recover, that a worldwide boom would take shape, and that the industrial capacity and expertise that the company had built up put it in a position to benefit.

Their interest was not confined to Massey. They were prepared to borrow heavily and systematically to build up their holdings in a number of industries which they felt would prosper. One year before the war's end, they combined forces to buy a comparatively small company, Standard Chemical Limited. The deal was an ingenious one. By offering shareholders a chance to get their hands on the company's surplus cash and accept a tax-free capital gain, they were able to purchase a company that was rich in assets for a bargain basement price. Shortly after this, they got together with a senior partner in Dominion Securities, J.A. (Bud) McDougald, and bought a large interest in a retail food firm, Dominion Stores, from a Frenchman who was anxious to repatriate a sizable sum in Canadian dollars to Europe.*

To finance his share of the Standard Chemical takeover, Taylor had been forced to borrow over $500,000. Even so, he was able to get more bank financing and continued to buy Massey stock. It was a pattern that was to become an established one, and its form and purpose were to be crystallized in an investment firm Taylor incorporated in Ontario in September, 1945, Argus Corporation.

The inspiration for Argus was a U.S. company named Atlas Corporation, which was put together by a successful American financier, Floyd Odlum. Odlum knew Taylor. He would advise Taylor and the other Argus' partners, Colonel Phillips and a lawyer named Wallace McCutcheon. McCutcheon had been put on the board of Canadian Breweries in prewar days by a bank that wanted him to keep watch on Taylor and his expansion plans. In the process, Taylor and McCutcheon had become friends. On the fringe of the Argus group, but remaining outside it for the time being, was Bud McDougald. He had reservations about an investment trust like Argus, arguing that the shares in such companies usually sold at a discount from their net book-value. But he agreed to participate in another company, Taylor, McDougald and Co., which would put together deals and take an interest in them.

Floyd Odlum not only counselled the Canadians, he also had one of his vice-presidents, Roger Gilbert, installed as a director of Argus. Odlum himself had started from humble beginnings, as a lawyer for a public utility company, Electric Bond and Share Company. He had branched out on his own and had greatly benefited from the stock market crash of 1929, buying selectively and cheaply when others were selling, and going on to construct an investment-based business empire.†

* French financier François Dupré had agreed to sell his controlling position in Dominion to Safeway Stores of Oakland, California, but held off at the request of the firm's Canadian president, J.H. William Horsey. Horsey approached McDougald who, within a matter of hours, undertook to raise $700,000 to keep ownership of Dominion Stores in Canada. Taylor and Phillips joined him to buy out Dupré and later incorporated their Dominion interest in Argus.

† The original investment company that Odlum formed was called the United States Company. Odlum, a lawyer friend, George Howard, and their wives, each put up $10,000. Later they were joined by other friends. and by 1928 their company had investments worth $600,000. By the following year, Odlum had registered Atlas in the State of Delaware, and was using his own and other people's money to buy cheap stocks following the crash on Wall Street. (At the time it was noted that the biggest buyers of depressed stocks were Floyd Odlum and Joseph Kennedy, founder of the Kennedy clan.) By the spring of 1930, Atlas had assets of $17 million and was set for further expansion.

In the 1930s, Odlum had seen that the market value of securities of investment companies was well below the actual value of the assets owned by the holders of these securities. By buying investment companies, he could acquire their portfolios at less than their actual value. He, had proceeded to buy no less than twenty-one investment companies including some big-name firms such as Goldman Sachs Trading Corporation, Shenandoah Corporation, and Blue Ridge Corporation, multiplying the holdings of Atlas several times over.

The approach Taylor and Phillips were to take with Argus was different than Odlum's to some degree. They were bent on achieving hegemony over a few major Canadian companies. Their investment company would put the greater part of its funds into a few firms that were judged to have great potential for the future. Argus would choose the companies on the basis that it could, with anything between 15 percent and 35 percent ownership, become the largest individual shareholder. The purchase of the shares would be financed through increased bank borrowing. When it was accomplished, the Argus principals would press for a significant representation on the board and the formation of an Executive Committee of the Board which they would control. Argus would grow by reinvestment. In the early years, it would put aside the major share of its cash income so as to strengthen its position and to acquire new shareholdings.

Taylor lumped his major interests, including Canadian Breweries, Dominion Stores, Standard Chemical, Caninvesco, and Invesco, into Argus in return for 50 percent of the issued shares. This included the major part of his Massey-Harris holdings. Phillips sold his Standard Chemical stock to Argus while McCutcheon paid cash for his interest. The result of this amalgam was a partnership in which Taylor had the greatest equity interest. Despite this, Argus was run and organized as a three-man effort with Taylor going to considerable pains to consult with his business partners. In some areas, they could contribute more than he did. Phillips as the administrator and adviser could counsel companies on management problems; McCutcheon combined a keen intellect and an ability to get things done with ruthlessness when needed. But there was no doubt where the company's entrepreneurial spark came from. It was E.P. Taylor, with Bud McDougald, who could put the deals together and provide the forward momentum.

The companies Argus invested in would remain much the same outwardly, while the Argus directors and Argus-influenced directors would seek to further their expansion from inside. They might help with mergers or acquisitions or new financings, always playing a supporting role. In this way, Argus would grow as the companies themselves expanded, and as their net worth and earnings increased. It was not to be a passive participation. Argus would concentrate its efforts on certain companies and expect to be rewarded. Initially the companies chosen were Standard Chemical, Dominion Stores, Canadian Breweries, Orange Crush, Canadian Food Products, British Columbia Forest Products and, last but by no means least, Massey-Harris.

In the immediate postwar period, Taylor had directed his energies to organizing two companies, British Columbia Forest Products and Dominion Tar or Domtar (which was merged with Standard Chemical). But by the winter of 1947, he was free of these encumbrances and began to give his attention to Massey. At the annual meeting in February of that year, Taylor and Phillips were joined on the board by three other Argus nominees, Joseph Simard, John Tory, Q.C., and Harry Carmichael.* Through judicious behind-the-scenes purchases, Argus had acquired 150,000 voting shares out of a total of 1.1 million issued and outstanding — hardly a convincing or controlling position. And while Argus did have five seats on the board, it could not count on its wishes prevailing as a majority of seven seats were still in other hands.

Even in these early years, conflicts were developing. Duncan was plainly alarmed by the aggrandizement that had gone on and the position that Argus had accumulated. He recognized that the objectives of the Argus group were different from his own; they saw Massey-Harris as an investment, and a good one, and were not motivated by any particular feelings of loyalty, or of continuity and tradition. Duncan accused the group of acting as buyers of company stock on behalf of an American industrialist, Victor Emanuel.† And he had a quarrel

---

* Joseph Simard's close Liberal party ties had enabled him to buy the federal shipyards at Sorel, Quebec, before the war, and to make a fortune out of contracts to build navy ships. He was a friend of Phillips. John Tory was the most influential corporate lawyer of his day and closely associated with the Argus group. Harry Carmichael had been a dollar-a-year man and had taken Taylor's job in Ottawa when the latter moved to Washington.

† Temporarily short of funds, Taylor and Phillips had persuaded Victor Emanuel to purchase shares on their behalf. Emanuel was the president of Avco Manufacturing, a U.S. company that ranked as the third-largest in defence work. During the war, Avco had built 33,000 airplanes, 9 aircraft carriers, the battleship *South Dakota*, and bodies for nearly all the jeeps used by the armed forces. When the war ended, Emanuel was keen to use the firm's rich profits from defence work to diversify, and he launched the company into appliances, broadcasting, farm equipment, and laundry equipment, while retaining aircraft engine manufacturing as Avco's primary business. His farm equipment investment was a small U.S. company, New Idea Corporation, bought in 1945. Taylor and Phillips persuaded him that a merger with Massey-Harris might be arranged. If he was to buy in, then his stake and that of Argus would give them working control. The arrangement fell apart because of Duncan's opposition. One of the conditions had been that Duncan would continue as president of the amalgamated companies. He refused to do this, and Taylor and Phillips backed off. Later Emanuel sold Avco's shares to Argus, raising the Argus position in Massey to 14.4 percent.

with them in the Ritz Carlton Hotel in Montreal after Phillips
proposed that the aging Harry Gundy — who happened to
be staying in a suite in the hotel — should be asked to
resign. Phillips suggested that Duncan should tell Gundy of this
immediately, but Duncan would have nothing to do with it.

Duncan may have disliked what was taking place, but he
was powerless to stop it. In any case, he felt he could maintain his
own position and safeguard the independence of the company.
However, while he was an influential figure, he did not have
the kinds of contacts or the inclination to make deals and pull
the levers of power needed to ensure survival. In his own defence,
Duncan refused to develop any kind of strategy. That was not
his way of doing things and to resort to it would have been,
in his eyes, demeaning. Instead, he hoped to override Phillips
and Taylor with his years of seniority and experience in the
business. They needed him, he reasoned, while he had no great
need of them.

Early in 1948, a sixth Argus nominee came on to the board,
Wallace McCutcheon. The next move after that was the creation,
with Duncan's tacit approval, of an executive committee of the
board. The committee was formed in the same year. A small
cabal of board members including Taylor and Phillips, it met
every month to discuss the affairs of the company, intervening
and directing policy in a way that had not happened before.
Duncan remained at the helm, travelling around the world, an
ambassadorial figure delighting in his social contacts with
famous and influential people. But back in Toronto, real power
rested with the Taylor group which controlled the executive
committee and, after the election of yet another Argus camp
follower, Bud McDougald, the board of directors as well. Argus
with its minority interest had achieved a pre-emminent position
in the affairs of Massey, something no group of shareholders
had done since the Massey family sold out in 1927.

The Argus triumph had been accomplished by following
the strategy Taylor had been taught by Floyd Odlum, and
which he summarized succinctly in the following way: "I look
for companies where no very large shareholder exists. With
my partners, I buy enough stock to give us effective control.
Then the company holds our view."

Part Two:
The Duncan Years (1927-1956)

7

"A Many-sided
and Able Man"

*Restoration of Massey's position in Britain and
France, together with the eventual rehabilitation of its German operations, were to be the
foundation for its international business....To
Duncan it was a manifestation of Massey's
purpose and his own feeling for history.*

During the war years, Massey had been decisively revived
and brought back from the brink, with much of its recuperation
coming from rich contracts for war equipment to aid the Allied
war effort. Plants had been reorganized and re-equipped. New
factories had been opened. Huge assembly lines had turned out an
arsenal of war products, including the M5 and M24 tanks in the
U.S., and wings for the Mosquito fighter bomber in Canada.

All this had greatly changed the company from the struggling,
debt-ridden farm equipment maker that had barely avoided
bankruptcy in the 1930s. Before the outbreak of the war,
worldwide sales had been just $21 million. By 1945, sales had
climbed to $116 million, of which a record $68 million came
from war production. As Massey expanded during the war
years, so the character of the company's organization had
change. Equally far-reaching changes had taken place in the
markets which had constituted the biggest share of Massey's
business in the past. These markets were in countries financially
crippled as a result of the war. Even when peace returned,
Europeans, and to a lesser extent South Americans and
Australians, would have difficulty reorganizing their economic
priorities, and building up their farm industries and food
production. Further, those countries saddled with huge debts
could not be expected to commit scarce dollar reserves to buying
Canadian-and American-made farm machinery.

Within Massey, there was a sense of disquiet about the

company's postwar future. Could Massey continue to keep its plants open, and employ a work force that had more than doubled during the war years? Could it obtain the necessary raw materials for a major push in farm machinery production? Where would it find its markets? And how would it fare against its U.S. competitors? Massey's senior managers, particularly the Depression veterans, viewed global developments with extreme pessimism.

In Canada Massey's position was challenged only by International Harvester, based in Hamilton, Ontario.* International Harvester had moved to widen its sales into heavy duty equipment and trucks. Still, in Canadian terms, Massey was ahead, with assets of $49 million. This edge enabled the company to plan new activities and to buy machine tools and expand and modernize plants. Early on, a commitment was made to build a new factory for making combines on the north side of King Street in Toronto. Capital expenditures were advanced from less than $1 million in 1944 to $3.3 million in 1945 and $4.4 million in 1946.

A planning committee set up by Duncan had recommended that, after the war, Massey should diversify into a totally new line of products, making refrigerators and household appliances for the North American market, thereby turning its back on war-ravaged Europe. This was never acted on. Massey was to remain a farm equipment company. However there was a general air of unease about the future.

Although a big company by Canadian standards, in comparison with its American rivals — with which it would have to do battle in virtually the only affluent farm market left in the world — Massey was small and unimportant. There were at least five U.S. farm companies that were bigger than Massey.† Massey's assets of $45 million were hardly a match for International Harvester's $558 million and John Deere's $200 million.

None of this, however, was to dissuade Duncan from pursuing an ambitious course for the company, or dampen his enthusiasm for what he saw as the opportunities that lay ahead. Massey had done well with its war work. It had built up its name and reputation. It had been the first farm machinery

---

* International Harvester's founder was Cyrus Hall McCormick who, in 1831, invented the Virginia reaper, a machine that could cut as much grain in a day as fifteen men with sickles, thus claiming for himself the honor of being the father of farm mechanization. In 1902, five farm companies in the U.S. had merged to strengthen International Harvester and, in the following year, the company had opened a plant in Hamilton, Ontario. Employing 800 workers, it was built to serve the Canadian market. With its strong base in the U.S. and abroad, International Harvester has provided stiff competition for Massey ever since.

† International Harvester, John Deere, Case, Minneapolis-Moline, Oliver, and perhaps Cockshutt, were all ahead of Massey.

company in the world to put the self-propelled combine into mass production. Selling into the U.S. market had been helped by the removal of tariffs between the two countries. And although war production had constituted the major share of sales which, naturally, were bound to fall, farm machinery sales had moved ahead to $50 million a year. In the food-short postwar world, there seemed hope for even larger sales in North America. The farmers of the U.S. Mid-west and the Canadian Prairies would have to take on the job of feeding the world.

The key to the future was the U.S. where Massey faced the stiffest competition. Duncan had some confidence that the situation would improve in the U.S. but this belief stemmed more from faith than experience because Duncan still did not know the market well. His interests, and his upbringing, had been European, and he was not predisposed to visit the U.S. or take great interest in its political and economic affairs. Duncan's energies were to be concentrated on Europe and on creating a new Massey-Harris organization out of the ashes of war. In the U.S., he would support his deputy Ken Hyslop, and hope that things turned out for the best.

Fortunately for Duncan, and for Massey, the momentum the company had built up in the U.S. during the war was carried forward. And it was done so with imagination and skill, not by the languid Ken Hyslop but by his peppery sales chief, Joseph Tucker, known to his colleagues as Marshal Joe.

Joe Tucker's career had begun in the copper mines of Michigan. He then went to work for a small manufacturing company that made threshers. He had moved to a larger firm, the Oliver Corporation, in the 1930s before being recruited by Massey where he was vice-president in charge of U.S. sales. A lean granite-faced man, in his mid-fifties, Tucker lived in Evanston, Illinois. It was at his home in Illinois that he first came up with an idea that was to launch Massey into its greatest sales drive in the U.S.

In the fall of 1943, only a few months after joining Massey, Tucker received a letter from his daughter in China. She was married to a missionary and wrote to him about the deprivation and food shortages in the Chinese countryside. Food had been so scarce in their own family that, with her son, she had undertaken a risky thirty-mile trip down the Yellow River to the town of Fowyang. She had made the voyage, she reported, after hearing that there was a small stock of dry wheat in the town. "Luckily there was a small supply," she wrote her father, "enough to see us through the winter. Even though the grain was filled with dirt and weevils, even though I would have to grind it by hand, still we would have food."

To Tucker, sitting in his comfortable home in Evanston,

the plight of his daughter and the people of China was something more than a distant concern. Newly arrived at Massey, he had been pondering how the Canadian-based firm could make a greater impact in the U.S. corn belt. Now, this letter, with its report of desperate conditions elsewhere in the world, gave him an idea. Within a matter of days, Tucker, brandishing his daughter's letter, was in Washington talking to officials of the wartime Food Administration. The ending of the war in Europe and the Pacific would not bring an end to the miseries there. It would be the duty of the victors, Tucker said, to sustain millions of people who would be hungry and in need. What was urgently required was a plan that would raise the productivity of the North American food industry to meet this challenge.

Speaking on behalf of Massey, Tucker had such a plan. The firm that had pioneered the self-propelled combine would be prepared to use 500 of them to help harvest the following year's grain crop. If the U.S. and Canadian governments would give their approval, and allow the release of materials to build extra combines, then Massey would undertake to sell the machines only to operators who would promise to harvest at least 2,000 acres each. If the operators did not have this amount of land themselves, they would be hired to work other land. In all, the Massey-Harris Harvest Brigade would cover one million acres.

To the governments in Washington and Ottawa, the plan was appealing. Tucker's salesmanship was to make it still more appealing. By using self-propelled combines instead of a tractor-combine team, less manpower (one man instead of two) would be needed. A saving in fuel — 500,000 gallons, Tucker guaranteed — would be made. And the harvest would gather in a larger crop (Tucker calculated the additional yield to be 500,000 bushels). Persuaded by the scheme, the food planners gave their approval to it in both countries. Finding the materials to build the combines was up to Massey.

At this point, Tucker began to earn his reputation as Marshal Joe. Utilizing some of the expertise the company had acquired as a supplier of war components and products, and his own abilities as a scavenger and organizer, Tucker managed to get his materials and his Harvest Brigade.

In the spring of 1944, the big red machines started their work in the south. In April, an operator in Texas reported cutting a record acreage of flax. It was a rehearsal for the great wheat-harvesting to come, and a preliminary event before the machines, with their attendant publicity, moved north. By June, the whole brigade was at work on the plains. Advance crews would go into farm communities and find out where fields could be cut; repair crews were stationed close by; and an occasional light airplane would set down on the stubble with a batch of orders from Tucker to the men in the field. The machines worked alone, in pairs, or sometimes with as many as eight or nine in a row. They worked during the summer days and by floodlight at night; travelling north through Oklahoma, Kansas, Nebraska, and the Dakotas, towards the border and into Canada.

By September the combine brigade had crossed into Canada and the operation had been completed. Tucker had done everything he had promised — and he had decisively demonstrated the superiority of the Massey combine. As a result of the drive, production had risen to 1,800 machines in 1944, up from 760 in the previous year. It was a real bonus for the firm. More importantly, Massey had won a battle on hitherto rough terrain. The U.S. had been the scene of several defeats, and no clear cut victories. Now the firm had come out ahead of its competitors in a single area, though it remained outgunned and vulnerable in others. In self-propelled combines, U.S. Mid-west farmers had come to realize that the Canadian firm had no serious rivals.*

"A terrier quivering with eagerness."

The achievement of the Harvest Brigade had been a brilliant marketing coup. It had also pointed out to Massey head office that, notwithstanding the devastation caused by war, the time was ripe with opportunities, and that an original approach and good contacts at the government level could create fresh business where none had existed before.

In the immediate postwar period the board of directors seemed to forget the lesson of the Harvest Brigade. As they looked around the world, they were captivated by the bright prospects that seemed to be offered in the U.S. and Canada but doubtful about the wisdom of investing time and effort elsewhere. Duncan needed little convincing. He knew Europe would recover swiftly, and he was adamant that Massey must devise a strategy for getting back into Europe in a major way or lose its position as a front-rank international company.

In the company's first annual report after the war, Duncan informed shareholders that "an unprecedented demand" existed

---

* By 1948 Massey had captured 53 percent of the total U.S. market for self-propelled combines. By contrast, it only had 3 percent of the market for tractors and, despite the success of the combine, its overall share of the farm equipment market in the U.S. was an extremely modest 4 percent.

for farm goods and machinery and that this would "far exceed the capacity of (the company's) manufacturing facilities in 1946." There was indeed a discernible increase in demand. Governments had removed many of the restrictions on production. U.S. and Canadian markets were turning up, while in Europe, the postwar black market had helped to enrich rural families who were keen to replace farm equipment that had been destroyed during the war. Still, the opportunities that Duncan perceived had to be exploited with skill and caution. It was a question of negotiating with governments and responding to their wishes rather than head-on rivalry in the marketplace.

This was an area in which Duncan excelled. He could scarcely wait to quit his King Street office and get started on his travels to seek out the business opportunities he knew were waiting. A reporter from *Fortune* magazine who did an article on Massey-Harris after the success of the Harvest Brigade quoted a description of Duncan from one of his colleagues: he was like "a terrier quivering with eagerness."

In Europe, Duncan uncovered tragedies as well as triumphs. On arriving in Paris he learned that the European general manager who had taken over from Ken Hyslop in 1938, a Latvian Jew named Sigismund Voss, had fallen into the hands of the Gestapo and been taken to the death camps. Voss had insisted on staying at his post in Paris until he was arrested.* In London, Massey offices had been hit by German bombs. In France, the plant at Marquette les Lille had suffered damage from British bombers. In Germany, which Duncan obtained a permit to enter only weeks after the end of war, the scene was of almost total destruction. The plant at Westhoven had been obliterated in night air raids. It had been rebuilt and converted to manufacture weapons, only to be destroyed once again. The walls were shattered, and the floor piled high with masonry. The only recognizable memento was a plaque that read: "Erected to the memory of our glorious dead, Massey-Harris fellow workers, who died in defence of the Fatherland."

The physical destruction was substantial all over Europe. More than this, Massey had lost many of its personnel. They had been dispersed or had perished during the war. To compound these problems, there was the economic plight of the various combatants; even in victorious Britain the economy was in tatters. While the British might be in urgent need of farm

---

* Because he was Jewish, the Germans had insisted that Voss must be replaced by a French manager. Subsequently, he had sent his family out of the country to safety, but he insisted on going only to the city of Nantes, close to the free zone, and continued to manage Massey's affairs from there. He was arrested, transferred to the Fresnes prison outside Paris and transported to Auschwitz where he perished. While he was in prison, Massey-Harris had attempted to get him released and repatriated to Canada but in vain.

equipment to stimulate food production, it was clear they would not be buying tractors, hay-making machines, and combines from across the Atlantic. Instead they would have to make their own.

Rather than causing Duncan to plunge into despair, the incredible problems facing Massey in Europe seemed to him to hold out opportunities. Though its factories had been razed and its organization shattered, Massey could still reclaim its former strong position and become a partner in rebuilding Europe's farm economy. The board of directors and the Argus group might prefer to play it safe and gear production to the surefire prosperity of the U.S. but Duncan wanted to do things differently. He knew his way around Europe; he could open doors and make his influence felt. All that was needed was to have a word with the right people at the right time, and Massey would re-establish itself.

In London, Duncan's contacts and his friendship with Britain's political leaders did indeed open doors. In successive days he was able to see the minister of agriculture, the minister of supply, the permanent undersecretary of the Board of Trade, and the foreign secretary, Anthony Eden. All spoke highly of Massey's contribution to the war effort, and its help in shipping farm machinery to Britain. Massey was the largest supplier of combines and harvesting equipment in the country. But each of the top ministers — introduced to Duncan by the Canadian High Commissioner, Vincent Massey — was adamant that imports from North America would have to be curtailed. Britain did not have the dollars to pay for them. If Massey was prepared to set up a manufacturing plant in Britain it would be welcome and every assistance would be given. Otherwise, access to the British market would come to an end.

Duncan had never shied away from making major decisions. Almost instantly, he committed Massey to building a new plant and starting up manufacturing. It was a decision that might properly have been left to the board of directors, but Duncan was confident he would overcome any reluctance they might have. So he ordered a team of Canadian engineers to survey sites and find sources and supplies of raw materials. The final decision to go ahead would be delayed until the summer. But, in practice, it had already been made. Duncan had committed Massey to investing in Britain, an investment which was to have far-reaching consequences for the future.

Having made the commitment, and completed his tour, Duncan went to North Africa and from there he flew by military plane into newly liberated Paris. Le Bourget airport was still strewn with wrecked Nazi planes. The city had no heating; electricity was rationed; and transport consisted of horse-drawn buggies and bicycles. The Canadian ambassador, Georges

Vanier, (later to be Governor-General) had just returned to take up his post, and Duncan was his first visitor. Once again, doors were swung open. Duncan went to see the head of de Gaulle's cabinet, Gaston Palewski, and called on the military commander of Paris, and a top foreign affairs official. He then travelled north, in a car lent to him by the ambassador, journeying on rutted roads and detouring around bombed-out bridges. The Massey factory near Marquette les Lille had been damaged by bombs — twenty-six had fallen in and around the plant. But it was still operating, with a work force of 500 men who had continued to manufacture farm implements throughout the war.

Restoration of Massey's position in Britain and France, together with the eventual rehabilitation of its German operations, were to be the foundation for its revived international business. Duncan, with his impressive connections, was able to ensure that the company did not go unrecognized by the politicians and bureaucrats. If decisions were being made about investments, or steel, or materials allocation then, thanks to Duncan, Massey would be on the list and would receive aid and encouragement from the planners and officials; it would remain a company with a presence in Europe.

To Duncan the direction and thrust of these activities was clear. They were a manifestation of the ideals he had always held for Massey and his own sense of history. Massey had thrived as an international firm. It had faced trouble when, in the 1930s, temporary economic setbacks had prompted those in charge to draw back, to lose the sense of vision of men like Hart Massey and his own father.

Such considerations did not, however, loom large back in Toronto. The only equations that appealed to E.P. Taylor and Eric Phillips in their extended financial position were those of profit and loss, and while it was apparent that Americans and Canadians needed farm machinery, and could afford to pay for it, the same did not seem to be true of Europeans. What Duncan was proposing was to them a risky undertaking. And while they were prepared to go along with it for the time being, no major investment would be made. When Duncan told them about the planned factory to be located in Manchester, Taylor and Phillips made sure that the board voted a minimal $500,000 for the project. As for aid to rebuild the French and German plants, they vetoed the idea altogether.

These refusals were not in fact as damaging as they might have been. The French company was able to climb back because of the demand for its products, even though it lacked new capital equipment. The German operation remained at a subsistence level — as late as 1948 Massey's German workers were receiving food parcels to keep them going. But it, too, would eventually recover.

Meanwhile in Britain, Massey was being drawn into making a large commitment, but not one that was to entail a big financial expense. In Manchester, a war factory was leased from the government. Financing initially came from the $500,000 bank loan that Taylor and Phillips had permitted, while head office extended other credit in the form of materials, machinery, and implements. This outlay was recouped later through local financing and public stock issues. The investment had been made cheaply and Massey had retained ownership and control.

An even better deal was to come along a few years later. Britain had a desperate need for edible fats, which it wanted to import from the sterling area rather than from dollar countries. So the proposal was hatched by the United Africa Company, a subsidiary of Unilever, to create a huge 2.6 million acre ground nut plantation on the high plains of Tanganyika (Tanzania). The so-called ground nut development scheme was taken up enthusiastically by the Labor government of Clement Attlee. Scarce foreign exchange was found to buy farm machinery from Massey in Canada, and it was made clear by the scheme's sponsors that even more valuable contracts could be obtained if the company manufactured tractors in Britain.

The British government then offered to construct the desired factory in Kilmarnock Scotland, to be leased back to Massey at a rent of just $30,000 a year. With its contract for the ground nut scheme lined up, and new facilities provided, Massey was now ready to expand its sales substantially in Britain, and in Europe, and to do so courtesy of the British taxpayers.

As it turned out, the site chosen for the ground nut plantation — an arid section of the Tanganyika plain — proved totally unsuitable. After the expenditure of millions of dollars in public funds, the scheme dissolved into a scandal that was to permanently damage the reputation of the Labor government, and lead to its eventual replacement by Winston Churchill's Conservatives. The grand political fiasco did not injure Massey's interests. The company was now producing a full range of farm equipment in Britain, where it had established itself as a major manufacturer both quickly and cheaply. As an added spinoff from the ground nut affair, Massey had set up distributorships in Africa.

Expansion was occuring elsewhere in Europe during the late 1940s. After the first harsh years of cold and hunger, the nations of Europe now had access to aid money from America with which to build up their depleted reserves of food, and put agriculture on a sound basis. The plant at Marquette les Lille in France was expanded with government help and began to manufacture the small Pony tractors so ideal for French farms, and large self-propelled combines. Both the tractor and the combine were phenomenally successful in France, turning Cie Massey-Harris S.A. into a company with a nationwide reputation. In Germany, after a slow start, the Westhoven plant began making roller chain under contract to the British factories, and then moved into farm implements and combines. To encourage the company to provide jobs in the region, the state government of Hesse donated land and buildings at Eschwege. Once again, at minimal cost, a new factory was established.

If there was a pattern to Massey's postwar expansion in Europe, it was one in which the company benefited from the largesse of governments anxious to develop their farm sector. And in postwar Europe, buoyed by the infusion of U.S. aid and the knowledge that things could get better, opportunities had not been hard to find.

A napoleonic view of things.

On March 28th, 1947, politicians and business leaders from Canada and the U.S. gathered in the Banquet Hall of the Royal York Hotel in Toronto to celebrate the one-hundredth anniversary of Massey-Harris. There were 700 people present; the speeches were broadcast across the country on the CBC network, and the evening's festivities ended with the showing of the film *A Romance of Two Hemispheres*. During this gala event, fulsome tributes were paid to James Duncan. He was described by the guest speaker, Leonard Brockington, as "a many-sided and able man."

The description was apt since Duncan combined great talents with a remarkable ability to accomplish objectives. He had always had an enlarged view of what could be accomplished, and his energy and vigor had contributed a great deal to Massey's diversity, the nimbleness with which it had exploited new inventions and new markets, and the relative ease of its shift from wartime to peacetime operations. To a great extent, the many sides of Massey were a reflection of the many sides of James Duncan. Still, Duncan as a manager and administrator did have serious shortcomings. Like many able men, he could be wilful and egocentric. And with his talents and abilities, he was extremely vain.

At Massey, these flaws were personified in his almost regal approach to the presidency and chairmanship. All power resided with Duncan and all decisions were made by him. If there was travel to be done, it would be done by Duncan. If opinions were

114 to be voiced publicly about Massey's present or future course, then those opinions would be Duncan's. To have a different view of things, as, for example, Taylor and Phillips had, was to be disloyal to him and his vision of the company. In the case of Massey directors, Duncan might have to tolerate such dissension, but within the company he ruled supreme. To disagree with Mr. Duncan (no one ever dared to call him Jimmy to his face) was tantamount to treachery.

This napoleonic view of things worked well at certain times. In matters that interested Duncan — the promotion of the self-propelled combine, or the reconstruction of a postwar presence in Europe — action was taken swiftly and successfully. But in others where he displayed less interest — for example, Massey's attempt to capitalize on the Harvest Brigade in the U.S. and emerge as a leading farm equipment supplier — the initiative had come from other sources and, because it had not come from him, little was done. A poor administator, he was not keen on putting any kind of management structure in place, or delegating authority. To do so would have hindered his own use of power. For James Duncan a great company, like an ocean liner, could have only one person at the helm.

Not surprisingly, these autocratic habits had an effect. The farm equipment industry, having been scorched by the Depression, tended to recruit and promote from within, and to rely on sales-oriented people who knew the implement business and who had often grown up on the farm and become farm machinery dealers. These were the people who filled the ranks of Massey's executive class in the 1940s and 1950s. They were not businessmen, nor were they professional managers. Nor, since Duncan allocated so much authority to himself, did they learn very much about either business management or the worldwide nature of the company they were engaged in running. Duncan made the decisions. His staff — with the exception of the top managers in the U.S. — merely carried them out.

In 1944, in preparation for the postwar period, Duncan devised a new organizational plan. The firm had become increasingly complex and difficult to administer centrally as it spread out its war work and dealt with hundreds of subcontractors. Seven vice-presidents — soon to become known as the seven dwarfs — were appointed. They reported to a first vice-president, the duck-shooting Ken Hyslop, and through him to Duncan. The system survived but it was very much a pro forma one.*

_____
* It was so pro forma that despite his elevation to the first vice-presidency Hyslop, who was to retire in 1950, continued to be based in the U.S.

In practice, Duncan delighted in ignoring lines of responsibility and confusing things. He would call in one vice-president and ask him to solve a problem, then half an hour later ask another to perform the same task. Neither man would know about the overlap unless they found out during the course of the assignment. But the arrangement satisfied Duncan. With two reports, or three or four, on the same subject, the final decision would be his to make.

Of the seven vice-presidents, four were the most prominent; Herb Bloom, vice-president in charge of sales, Ed Burgess, vice-president in charge of manufacturing, Guy Bevan, vice-president in charge of engineering and research, and Walter Lattman, vice-president in charge of supply and control. The four men were a study in contrasts, a diverse group of personalities, and a cross section of the patchwork that made up Massey-Harris. As individuals they had little in common, but as senior managers they were bound together by a sense of loyalty to the company, and to Duncan.

Herb Bloom had first met Duncan in the late 1930s. At that time Bloom was manager of the Saskatchewan branch of Massey-Harris. A farm boy from the west, he had the well-rounded affable appearance of a university professor. Duncan deferred to Bloom on matters involving sales, and regarded him as an expert on the affairs of the west, and on farm implements. When it came to farming on the Prairies, Bloom had the firsthand knowledge and experience that Duncan lacked. He was promoted to top sales positions in Canada and the U.S. But when it came to the choice of a potential successor Duncan always ruled him out: Herb Bloom was not an internationalist.

The same could also be said of Ed Burgess, a tool-room apprentice who had learnt his craft on the factory floor and who combined his great technical skills with a sharpness and cockiness that appealed to Duncan. Of all the vice-presidents, Burgess was the most influential, and the only one to risk crossing Duncan (though he never risked referring to him as anything else but Mr. Duncan). Often he would win Duncan over with his persuasiveness and his gall. Duncan would give in to Burgess, even while he held out against the saner, more experienced advice of Herb Bloom and Walter Lattman.

Lattman, a cultured and distinguished-looking Swiss, and Guy Bevan, a Cambridge-educated engineer, would seem to have corresponded more exactly to Duncan's own mould. Lattman spoke several languages and knew the business well. Bevan had useful social connections in England and Europe. But neither of them seemed to measure up to Duncan's ideal of the aggressive, self-made man. Perverse it may have been, since

Duncan did not come from humble origins himself, but his greatest regard was reserved for men who had their roots in the soil and in the farm and machinery business, people like Herb Bloom and Ed Burgess and another up-and-coming salesman from the west, Bill Mawhinney. Duncan was loyal to all his vice-presidents, and believed that each of them had something to contribute to his own decision making, but he was especially loyal to the westerners.

In the immediate postwar period, the one-man rule that existed at Massey, and the overall lack of management plans and structures, was no great handicap. Markets were expanding rapidly. And the farm machinery companies, even the big ones in the U.S., had yet to feel the impact of the managerial revolution; by and large, like Massey, they were led by the self-educated and conservative men who had come through the Depression. Apart from a few engineers, Massey had virtually no senior executives with university degrees. But then neither did John Deere nor International Harvester.

Changes were coming, though as yet they were in the formative stage. Shortly after the end of the war, while on a visit to the Weston aircraft factory, Duncan had asked the factory's manager, Ed Burgess, about the factory's employees. One of them, said Burgess, was "a bloody schoolmaster." There and then, a young man named Rik Kettle, who had taught at Upper Canada College for ten years, was presented to Duncan, and was immediately transferred to the company's headquarters staff. Kettle was given the job of attracting and hiring bright young men, with degrees from Harvard, or the London School of Economics, or the University of Toronto, who would become the managers of the future at Massey. Many were recruited, and went on to become successful. But a lot of prejudices had to be overcome first.*

The effort to attract a young group of university-educated recruits had Duncan's blessing. He had great faith in the future and did not share the forebodings of some of his vice-presidents whose vivid memories of the Depression kept them perpetually on the lookout for signs of dust bowls and hard times. To them, there was little point in any kind of corporate planning. The farm industry would run its zig-zag course no matter what was done. Duncan did not share this belief. Ever the optimist, he was convinced that markets would always bounce back and that troubled

---

* One of Duncan's westerners was amazed when he learnt how much was to be paid to a young man who had a university degree and fluency in several languages. Shouldn't a young man from university get less money, he wanted to know. How could his knowledge of the industry compare with that of someone coming out of a dealership or off a farm?

economic times could best be ridden out by companies committed to expansion.

This attitude explains his readiness to recruit a new stream of executives — he especially favored those who spoke foreign languages — while at the same time continuing to rely on his old-timers to learn what was happening in the marketplace and on the farm. In his personal style, too, Duncan cut a sometimes curious and contrasting figure. A socialite and city dweller, he delighted in putting on gaiters, a yellow sweater and a peak cap, and going out to the Massey farm outside Toronto to play the part of the country squire. Often he would receive distinguished visitors there and take them on a tour of the estate to see his prize bulls and cows. The farm had a practical purpose. It was used to test machinery. But it was also part of Duncan's own self-cultivated mystique, and his vanity. Photographers and portrait painters would be invited to picture him in these surroundings, a pipe in his mouth, a stout cane in his hand, chatting with a prominent politician or a foreign ambassador.

Personal vanity, fed by the approbation that he was accorded, had become part of Duncan's makeup, and an integral part of the way Massey was run. In theory, the company was adapting itself to the rapid changes that were occurring in the postwar world, and opening its ranks to a new generation of managers. In practice, Duncan ran the place as autocratically as a Bourbon king.

The Ferguson System

Harry Ferguson was born on November 4, 1884, the fourth child in a family that was to grow to eleven children, and grew up in the tiny village of Growell, about sixteen miles south of Belfast. His father was a prosperous farmer, a fierce anti-Catholic of Scots ancestry, and dour and intolerant with his children. No other book was permitted in the Ferguson home apart from the Bible.

To escape the oppressive atmosphere at home, young Harry considered emigrating to Canada. He was on the verge of buying a steamship ticket and leaving Ireland forever when his elder brother offered him a job in his car workshop on Belfast's Shankhill Road. Harry was eighteen, and he was fascinated by machines and motor cars. He decided to stay.

Over the next few years, Harry was to build a remarkable reputation. One year after beginning work in his brother's modest workshop he had built a motorcycle to his own design. He then became interested in the aviation industry and built a monoplane which flew successfully in several competitions. After crashing and injuring himself, he decided to give up flying. He then took up car racing, an activity that in 1912 was scarcely less dangerous than flying. He had by this time started his own dealership selling and repairing Vauxhall cars. While testing the steering of a three-litre works car that Vauxhall had given him to race, he skidded the car, turned it over and landed in a ditch. The accident took place on the eve of the Coupe de l'Automobile at Dieppe in France. Harry walked away from the crash. But Vauxhall were less than enthusiastic about losing one of their finest works cars, and decided their racing partnership with Ferguson was over.

These exploits were not the result of youthful high spirits or brashness. Harry Ferguson, self-taught as an engineer and designer, was a serious and purposeful young man; his 90-m.p.h. crash in Dieppe came when he braked hard to show a friend what he considered to be a serious defect in the steering of the car. When the First World War came, Ferguson was approached by the department of agriculture of the Irish government. They wanted to know if the young mechanic and inventor — who by this time was also selling farm tractors — would help increase farm mechanization and the use of tractors throughout Ireland.*

The choice of Ferguson for the job was an imaginative one. He immediately brought all his talents and inventiveness to bear on the problem. In his business, known as May Street Motors, he had recruited an enthusiastic engineer and designer, Willie Sands. The two of them toured Ireland as a team. Ferguson was the man of ideas. He thought of new concepts, and was able to demonstrate and sell them brilliantly, while Sands, who was the better engineer, would translate Ferguson's ideas into metal and machinery. By the end of the war, the two had come up with an idea that was to revolutionize agriculture. Instead of farmers having to hitch their ploughs to heavy tractors, an ungainly and inefficient method of ploughing, Ferguson conceived of a lightweight single system. The plough would become a unit with the tractor, making it far more stable and easier to steer.

The Ferguson System was to be developed and patented around the world. And it was to link the name of Harry

---

* British farms had lost their workers to the war effort while German U-boats sank much of the merchant shipping heading for British ports, thereby creating a serious food crisis. To ease the shortage, the government in London requested Ireland to convert pasture land to food production. The response was so great that, by the end of the war, Britain was receiving more food from Ireland than from any other country.

Ferguson with another great industrial pioneer, Henry Ford.
While building up his motor car empire at Dearborn, Michigan, Ford had turned his attention to the rich market for farm tractors, and was attempting to do something about their heaviness and lack of mobility. His solution was the light and agile Fordson which he agreed to supply to the British market as well as in the U.S. With the old hitch-on ploughs, the Fordson tended to become unstable and tip up in front. But the Ferguson plough could be designed especially to fit the Fordson.

Though the partnership was a natural one, it was to be excessively complicated because of the personalities of the two men, and eventually prove abortive to both of them.

Ferguson's first contacts had been with Ford's chief of staff, Charles Sorenson. When Ferguson arrived in Dearborn and demonstrated his plough, Henry Ford was impressed. But to him Ferguson was a small-timer from Ireland; certainly not a man to be seriously considered as a business partner. Turning imperiously to Sorenson, Ford had declared: "We could use a guy like that. Hire him, Charlie." He then turned around and stalked off. The embarrassed Sorenson was left with the job of trying to get Harry Ferguson on to the Ford payroll. He tried to entice Ferguson by raising the salary figure again and again, but he was met with continued cold silence. Ferguson would talk about the possibility of selling manufacturing rights to his plough to Ford but nothing else.

The first meeting had taken place in 1920. For the next nineteen years, Ford and Ferguson went their separate ways. Ferguson made arrangements with other smaller U.S. manufacturers, went to work perfecting his plough, and, in 1933, designed and built his own tractor. Ford stopped manufacturing tractors in the U.S. in 1928, though he still held out hopes that he could successfully return to the business later.

When the two of them were eventually brought together again, it was a confrontation of titans. Ferguson was by this time acknowledged as an inventor with a worldwide reputation, while Ford with his mass production methods had transformed the industrial landscape. This time, there were to be no misunderstandings. The two men, each irascible and independent were to take each other's measure as full and equal business partners. If anything, Ferguson gained the initiative in the encounter. He had always had great dreams for his Ferguson System and saw it as a concept that would advance mankind and create a new world of agricultural plenty. Nor was he reluctant to put his case in these terms. It was a world view which Ford shared and responded to, and by the end of their meeting the two men had agreed to work together. They would proceed to jointly develop a tractor designed for the Ferguson System, with Ford providing the product engineering and financing, and a joint company being set up to subcontract some of the work and handle distribution.

By the spring of 1939, the project was ready to go. Henry Ford was lavish in his praise of Ferguson. "His system will not only revolutionize agriculture," stated Ford, "but will put his name alongside those of Edison, Bell, and the Wright Brothers." Between two such mutual admirers and industrial pioneers, there was no need for contracts to be signed or documents to be drawn up. Their co-operation was marked by a handshake and nothing more. It was an agreement between gentlemen.

Manufacturing began at the end of the year. For the next eight years Ford was the maker and supplier of a tractor that was marketed by Ferguson under the dual name of the Ford tractor-Ferguson System. There were difficulties and restrictions during the war, but the arrangement generally worked splendidly for Harry Ferguson. His U.S. company saw sales rise from $5 million in 1939 to nearly $80 million by 1946. For Ford, however, the arrangement turned out to be far from satisfactory. The company was in a state of chaos that had nothing to do with its tractor venture. After the death of his son Edsel, Henry Ford, now in his eighties, had returned to run the company, but had handed over control to an ex-sailor named Harry Bennet who brought in armed thugs and launched a reign of terror against Ford's workers and managers. Henry Ford II, Edsel's son, was eventually to persuade his grandfather to name him president and oust Bennet and his gangsters. But by that time the company was losing $10 million a month and in danger of collapse.

In these circumstances, all Ford's manufacturing operations were under review. Among them was the far-from-favorable gentleman's agreement under which the company manufactured tractors but had no control over selling them. Ford wanted to market a tractor for itself. The company therefore began to develop a tractor named the Ford 8N that was basically the same as the one it was already producing, ignoring the fact that patent rights were held by Harry Ferguson.

In June of 1944, the Ferguson officials in Detroit were telephoned by a Ford executive who asked them how long it would take them to clear out of the building and move across town. No such move could be made, they replied, without consulting their management. Back came the response: "If you don't get out right away, you'll find yourselves on the street." Though the existing arrangement was to last for a few more years, relations between both companies deteriorated fast. Pressure was put on Ferguson to sell a majority interest in Harry

Ferguson Incorporated to Ford. When this was resisted, Ford
prepared to launch its own tractor.

A report by *Time* magazine on July 21, 1947, described the glittering ceremony held to inaugurate Ford's tractor. The magazine quoted young Henry Ford as saying that "since the days of my grandfather, we have always had one foot in the soil and one foot in industry." The writer of the magazine cryptically noted that "the foot which Henry put in the soil last week also came down hard on the neck of an angular Irish inventor named Harry Ferguson." That angular Irish inventor was not a man to be trifled with. Ford had expected that Ferguson would sue for violation of patents. If so, it would be a very private suit. Instead, Ferguson hit back with something much more embarrassing, especially for a company engaged in work for the government under the Marshall plan. He launched a $251 million anti-trust suit, accusing Ford of monopolistic practices in conspiring to try and put him out of business.

Ford tried to settle out of court, and then delayed the proceedings. Ferguson gritted his teeth and stayed with it to the finish. When a journalist asked him whether he would be glad to see it all end, he declared that it had become a hobby. "Some people play golf. Others go fishing. I have my case." In fact, the publicity surrounding the case of the little Irish inventor taking on the mighty Ford Corporation was of no small help; it aided Ferguson's new manufacturing business in the U.S., (a factory had been built in Detroit and completed in the summer of 1948), as well as its growing business in Britain and Europe. When a settlement came, Ferguson had to accept $9.25 million and an agreement from Ford that it would not infringe the Ferguson patents. The settlement was not finalized until April, 1952.

During this long-running battle, Ferguson had been looking around for allies and possible helpmates, and he had considered Massey-Harris. The Canadian company had a reputation internationally and it was weak in the design and manufacture of tractors, an area where his firm was strong. Perhaps James Duncan could be persuaded to take on Ford with him and co-operate in building Ferguson tractors in the U.S. The results of a meeting between Ferguson and Duncan in 1947 were inconclusive. Duncan had other plans and felt the commitment was too risky, whereupon Ferguson went ahead and built his own factory in Detroit. Contact between the two had been established, however, and Ferguson, like others before him, had been impressed with Duncan.

A few years later, it was Massey's turn to make an approach. A downturn in its business in Europe in the 1950s had prompted the company to look into the possibility of manufacturing a combine for Ferguson in its Kilmarnock plant. A deal was drawn up with Ferguson's general manager at the company's head office in Coventry, England, and Duncan — who happened to be in Europe at the time — arrived to sign it. When he reached the office he was told the deal had been scrapped by Ferguson. In a long-distance phone call from Ireland, where he was recovering from an operation, Ferguson had said that further negotiations with Duncan would be conducted by himself, and no one else. Ferguson had since returned to Abbotswood, his estate near Stow-on-the-Wold. He had that morning dispatched his Rolls Royce to Coventry. It was waiting outside, and Duncan was invited to go to luncheon with him.

The million dollar coin.

Ferguson's country home, Abbotswood, had been purchased just after the war. It was an elegant stone-faced house, which had been remodelled by the great English architect, Lutyens, at the turn of the century, surrounded by parkland. A trout stream crossed the estate and beautifully kept gardens and lawns — maintained by the staff who were all from Ulster — encircled the big house.

For Ferguson, his home represented two things; first, a pastoral retreat and an outlet for his private eccentricities, second, a place where he could meet people and transact business. Many of his eccentricities were on display at Abbotswood. He wanted the trout stream diverted into an artificial lake with an island in the middle of it, but the lake had to be perfectly circular when viewed from his bedroom window. His staff worked for days on the project, but he was not satisfied. He complained that the swirls of current did not break evenly on either side of the island. He was fastidious about the grounds being kept clean. Not a single leaf was allowed to remain on the grass. In the fall, a team of gardeners had to catch the leaves as they fell. Ferguson also insisted that all the trees in the parkland should grow in a perfectly upright way. To achieve this, some of them had to be roped down.

Inside the house, things had to conform to Ferguson's idea of perfection. The laces on his shoes had to be threaded so that there was an equal amount of lace to be tied on each side; the sheets on his bed had to overhang exactly on both sides; and meals had to be served at the correct time. If things did not go right, Ferguson would chide his family and his staff. But he was not a frightening man for all his unreasonableness. One of his stipulations was that his staff should not be treated as servants or overworked. They were not just paid retainers, they were partners and loyal associates in the better world that Harry Ferguson was building for everyone.

Among Ferguson's quirks was one that he carried into his business, and this was the notion that in the more equitable world, which men of vision like himself — designers and engineers — were creating, the price of goods would have to fall rather than rise. This would be to the betterment of everyone, and lead to the defeat of the evil virus of communism. Ferguson had preached the concept to Roosevelt, and claimed to have converted him to his Price Reducing Scheme, otherwise known as the Ferguson Plan. And he had sermonized on the subject to Churchill. After their luncheon together at Chartwell, Churchill whispered to a friend that he found Ferguson a remarkable fellow, but he had a bit of a one-track mind.

Ferguson carried out his Price Reducing Scheme in his own business. He could have made better profits in the late 1940s and early 1950s, but his ambition was to reduce the price of the Ferguson tractor built in Coventry from just under £200 to £100. While markets were moving up in the postwar years, he could practise his theories with impunity. But by 1952 the best was over. Farm machinery sales were turning down. The Ferguson company with its low profit margins had failed to build up any reserves as a cushion. This was especially true in the U.S., where the new Detroit manufacturing plant was a high-cost operation. If a price war developed, Ferguson in the U.S. could not survive the competition from giants like Ford.*

All these things were on Harry Ferguson's mind when he invited Duncan to come to luncheon at Abbotswood. He had one other worry as well. Nearing his seventieth birthday, he had no children to hand his life's work on to, nor had he — with his arbitrary ways — encouraged successors to come forward within his own organization.

After an excellent luncheon, Ferguson invited Duncan to come and talk in the garden summerhouse. He told Duncan that the proposal for Massey to manufacture a Ferguson combine was "so much smaller than what I have in mind that we should just forget about it." He then offered Duncan a 50 percent partnership in his company on condition that he resign from Massey. Duncan was taken aback. He refused courteously. Whereupon Ferguson lost his temper. "You'll

* Between 1948, when tractor assembly was started in Detroit, and 1951, Ferguson sales in the U.S. rose steadily from $33 million to $65 million, but the extraordinarily low margins that Ferguson took under his Price Reducing Scheme meant the company lost money in 1950 and never made more than $500,000 profit in the other three years.

124 regret it all your life," he told him. They then strolled around the garden until Ferguson, having regained his self-control, invited Duncan to resume their discussion in the summerhouse.

"Well, since you are not interested in a partnership, how would it be if our companies got together? I might be willing to sell some of my interests to Massey-Harris."

This new proposal at once interested Duncan, who was aware of Ferguson's strengths in the European tractor market. But he sensed that he might be being caught in a trap. On what terms would Ferguson be prepared to deal? he wanted to know.

Ferguson said he would start off by selling the U.S. business and, after that, Massey could buy into his British and worldwide interests over a number of years. It was what Duncan had feared. He replied that Ferguson's business in the U.S. was in a dilapidated condition and in trouble. He could hard recommend to his directors that Massey should purchase a company which was unprofitable, and showed little sign of becoming so. Ferguson once again told him he was making a mistake. And then they parted, with Duncan being told that once back in North America he should make a tour of Harry Ferguson Incorporated in Detroit. Perhaps it would change his mind.

Duncan did indeed make a visit. He also followed up on his discussions with Ferguson. The two companies could derive great benefits from a merger, he told the board, providing it was a worldwide one. The engineering and quality of Ferguson-made implements were superior and, by Duncan's estimate, if the companies combined forces, sales would rise at least another $100 million over the next few years. Duncan wrote to Ferguson saying the company might be interested in "co-ordinating its efforts with you in North America" providing this was the "first step towards a still closer union" worldwide.

That summer, Ferguson extended another invitation to Duncan to come to Abbotswood. He wanted to show him the new large tractor that he had been developing, known as the big Fergie. The cost of putting this tractor into production was one of the reasons that Ferguson was interested in a merger.

Before departing for England, Duncan worked out a number of possible deals. He had to be prepared for any eventuality since he did not know what the mercurial Ferguson might propose next. He took along the company's comptroller, Harry Metcalfe. When Ferguson and Duncan met at Abbotswood, luncheon was served — but just to the two men and their wives. Metcalfe and a senior Ferguson manager, Horace D'Angelo, were excluded; Ferguson desired only to talk to the top man. Luncheon over, Ferguson and Duncan adjourned to the summerhouse. Ferguson was direct: "I am prepared to sell you my company. What conditions do you have to offer me?"

Duncan was elated. However, he responded cautiously, saying that he felt others should be present, and the talks should take place more formally. Ferguson waved this aside: "A gentleman's agreement is quite good enough for me. Sit down now and tell me the conditions."

Duncan then outlined the Massey offer which involved a share swap. Seated opposite him in the summerhouse, Ferguson made notes on the back of an envelope. When he had finished, he was businesslike and brusque. "That seems fair. I agree with it. And I'll sell you my company on that basis." Duncan was astonished. He had been prepared to haggle over the details, even to offer Ferguson more money.

After the deal had been struck, officials from the two companies were invited into the summerhouse to go over the details. D'Angelo of Ferguson was silent when he heard about it. Like other company men, he had not been consulted or told in advance. Ferguson, as usual, had acted for himself.

The bitterness, and sense of loss, felt by the Ferguson men was to last for several years, and to sour relations with Massey. But in the aftermath of Duncan's offer and Ferguson's acceptance, there was another piece of negotiation to be concluded, and one that would have serious repercussions almost immediately.

Ferguson said there were two points he wanted to raise. First, he would like to have a position of honor in the new company, and second, he wanted the deal to be presented as an amalgamation rather than a sale. Duncan was in a generous mood. While he knew of Ferguson's stormy reputation, he felt that he could afford to be gracious; he therefore offered him the chairman's job, with the proviso that he would function as a North American company chairman — presiding over the annual meeting but taking no active part in company affairs — rather than in the executive role that a chairman often has in British firms. He also had no difficulty in agreeing to present the sale as an amalgamation. The new company would be known as Massey-Harris-Ferguson.

On hearing this, Ferguson became emotional. He wiped tears away from his eyes. "Mr. Duncan, I knew from the time I first met you that you were the man I wanted to be associated with," he said.

Duncan was embarrassed and dismissed the matter in an offhand way. By this time, the talks had dragged on through the afternoon. It was 6:30 P.M. and Ferguson assured Duncan that he had time to get dressed and be down for dinner at 7:00 P.M.

Duncan stalled. "Well, look, I want to phone Canada to let the chairman of the executive committee know what I have done. I think that is only fair. And I'd like to have a bath because I travelled overnight on the plane, so would you mind making dinner at 7:30 P.M.?" Ferguson went into a state of shock at the suggestion. He muttered that he always dined at the same time. He then agreed to try to change the schedule but came back shortly and announced that it could not be done. Duncan then rushed through his phone call to Canada, put his dinner jacket on, and came downstairs with his wife in time for a predinner glass of sherry. Shortly afterwards, the butler came into the room. There had been some trouble with the arrangements in the kitchen, he told Ferguson, and dinner would be delayed.

In the next few days, a group of Massey directors including Eric Phillips, Bud McDougald, Wallace McCutcheon, and John Tory, flew in and stayed at the nearby Welcombe Inn. They had every reason to feel happy with the transaction. Massey had acquired all the manufacturing rights and patents to the world's best-engineered tractor, together with an excellent staff of design engineers and a distribution system that put the company in first place in Britain, the Scandinavian countries, Australia and New Zealand, and several other countries. Harry Ferguson was to receive some 1,805,055 Massey shares which were valued at just under $9 each.

The price of the Ferguson acquisitions and its worldwide reputation was a cheap one — $16 million. What was also involved was the price to be paid for good will and expertise. At the last minute, one of his advisers prevailed on Ferguson to push his price up to $17 million.

This caused a problem. In the car on the way to a Ferguson tractor demonstration, three Massey representatives, Duncan, Phillips, and McDougald, engaged in an argument with Ferguson on the issue. The car had stopped in the picturesque village of Broadway when Ferguson proposed that the whole thing be settled with the toss of a coin. If he won, $17 million would be the price. If he lost, $16 million. The Massey people balked at first. It would be difficult to explain to shareholders a $1 million decision decided on the toss of a coin, but Phillips, a lover of sports and gambling, approved.

There and then, Ferguson borrowed half a crown from his financial adviser John Turner. The coin was spun in the street outside the Lygon Arms Hotel. Ferguson called tails, and lost. He then invited them to toss again, saying that if he won the coin would be his (though he'd already lost his $1 million). Ferguson won the second time and was about to return the coin to Turner when McDougald asked if he could keep it as a souvenir. The Massey people then took the coin to Asprey's, the London jewellers, and had it mounted on a silver cigar box. Inscribed on the box was "The $1,000,000 Coin" with a dedication underneath: "To our friend and partner Harry Ferguson. A gallant sportsman".

Ferguson bows out.

The unification of the two great farm empires of Massey-Harris in Canada and Ferguson in Britain was a momentous event. And it was presented as such. The press release spoke of the coming together of Ferguson which had "blazed a new trail throughout the world" and Massey which had "pioneered the self-propelled combine in every country where wheat is grown" as nothing less than epochal. It was "probably the most important news in the farm equipment industry in the present century."

Important and significant though it was, the merger had not come at a propitious time. Both companies were suffering from the post-1952 downturn in business, and sales were faltering. It was scarcely an ideal time to push through with a complicated merger. Moreover, no serious thought had been given as to how the two firms, with their competing product lines and over-lapping distribution systems, were to be integrated. In talking Harry Ferguson into selling out, and in winning his confidence, Duncan seemed to have scored a tremendous breakthrough. But the deal remained an illusory success, and was to create great problems for Massey for a number of years. Moreover, the sale had been made without any clarification of just what Harry Ferguson's role in the new firm of Massey-Harris-Ferguson was actually going to be.

It soon became apparent that Ferguson had a very different idea of his function from the one that Duncan had conceived for him. The friction started within a few weeks of the deal being signed. Duncan, invited with his family to spend a weekend at Abbotswood, was greeted in the hallway by a poker-faced Harry Ferguson: "Jimmy, I have heard from some of my previous employees that you called a distributors' meeting in London and I did not receive an invitation. How could such an oversight have taken place?" Duncan replied that there had been no oversight. As chairman of the board, the only meeting Ferguson would be expected to attend would be the annual meeting in Toronto. Ferguson disagreed. He felt he should attend every board meeting of a local sales company throughout the world.

The misunderstandings, once started, began to grow. Duncan had been warned by Sir Patrick Hennessy, the president of Ford in Britain, that Ferguson would be quarrelling with him within a matter of months. Hennessy cautioned that Ferguson had quarrelled with all his associates and partners and Duncan could not expect to be an exception.

The prediction proved accurate. Ferguson was concerned about the pricing of his tractors; about design changes; about the preference being given to Massey in developing and marketing the new model tractors. He wrote long letters to Duncan complaining about what was happening. At the end of a particularly bitter fifteen-page letter written in April, 1954 he declared that his relationship with Henry Ford Senior had been more satisfactory. "I never had to write him letters nor argue with him and waste precious time such as we have been wasting," Ferguson told Duncan, and he concluded: "If you will not accept my engineering recommendations then I will resign."

To be fair to Ferguson, there was some merit in his criticisms. Duncan had been showing too little concern over pricing under the contract that the company had with Standard Motors in Britain; and Ferguson had every right to be worried that giant competitors like Ford could undersell them. Still, Ferguson's constant interference was unwarranted. He seemed to want to direct the management of the company despite the fact that he had agreed he would not meddle.

To resolve the wrangling, Duncan and Bud McDougald flew to England to buy Ferguson out. Once there, however, they found themselves engaged in a cat-and-mouse game with an incorrigible old man. He said he had changed his mind about selling. He threatened to get back into the tractor business himself. He said he wanted an engineering staff from the firm to help him in his experimental work on a four-wheel-drive car. (An exchange with McDougald on this subject enraged Ferguson; after declaring that in his days with Henry Ford he had had a staff of 500 engineers and needed a similar number now, McDougald remarked casually: "We can forget it then, Harry. I had in mind two.")

The way the impasse was finally resolved was suitably esoteric. McDougald found out through an art dealer friend that Ferguson had made an appointment to see a painting by Constable that he wanted to buy. The appointment was for the following Thursday afternoon. Several frustrating weeks had passed and McDougald was still not sure whether Ferguson actually wanted to sell his interest or not. So he decided to call his bluff. Phoning John Turner, Ferguson's financial adviser, McDougald told him to get a message to Abbotswood to the effect that he had pressing affairs to settle in Canada; he would come to Ferguson's home on the following Thursday afternoon and endeavor to reach agreement with him, offering him cash

for his stock. But he could not delay further. On Friday, he
would be returning to Canada.

The ruse worked. On the following day, McDougald was able to find out from the art dealer that Ferguson had cancelled the appointment to view the Constable. He was serious about negotiating, after all.

When the talks did get under way, progress was slow. By the evening Ferguson was sullen and uncommunicative and insisted on the Massey team going off to dinner in the nearby town of Burford. When this happened, McDougald stayed behind to place a phone call to Toronto to make sure sufficient funds would be transferred if an agreement was reached. Coming out of a small sitting room, he heard Ferguson telling the butler that everyone had disappeared and that he intended to go to bed and call the talks off until tomorrow. McDougald argued with him, pointing out that he, Ferguson, had sent the Massey team off to Burford for dinner. Getting nowhere, McDougald jumped into a car and drove to Burford. Arriving in the hotel restaurant just as the soup was being served, he dragged Duncan and the other Massey people back to Abbotswood.

After this incident, Ferguson seemed more willing to come to terms. He had developed a great animosity for Duncan — his friend of one year earlier — and made rude comments about him. But, by 10:30 P.M., he had consented to his resignation from the chairmanship and agreed to accept the current market price for his shares. After weeks of indecision, and months of strife, Ferguson had finally agreed to bow out. But in his moment of departure, he could not resist making a final barbed comment about Duncan.

Drawing Bud McDougald aside, he told him: "You are a young fellow, Bud, but obviously you have got access to a lot of money. A fool and his money are soon parted and I would like to give you some advice. If you keep on backing this fellow [Duncan] with large sums of money, you are going to be out on the street." In view of what was to happen at Massey, Ferguson's parting shot was a telling one.

8

An Era Ends

Duncan was thought of as an indispensable man, by his staff, by the public, by his peers and, most of all, by himself. But among the people who mattered, the members of the Argus group on the board of directors, opinions of him were changing.

During one of his frequent trips to London, Duncan was the guest of the Soviet ambassaor to Great Britain, Jacob Malik. In the course of an amiable luncheon, Malik invited Duncan to visit the Soviet Union. It was during the period of the cold war, and Soviet invitations to western industrialists were rare. Duncan decided to accept. But he would do so only if the Russians would agree to his wife accompanying him on the trip; to three members of his staff coming along; and to the whole party being allowed to see whatever they wished in the Soviet Union.

The trip, and its outcome, were fully in keeping with Duncan's style and character. The Russians having agreed to his terms and conditions, the entourage set off in October, 1955. They toured the country in two chauffeur-driven cars while Duncan's aides made scrupulous notes about the journey, carried the boss's luggage, and conversed with the interpreters and drivers. Once back in Canada, Duncan wrote a series of articles which were published in newspapers across the country, at Duncan's urging, and put out in the form of a booklet, published and paid for by Massey-Harris-Ferguson.

The stories gave detailed information about life in Russia, including a comparison in dollars and roubles of the price of a bar of soap and cheap tickets to the Bolshoi Theatre (One can imagine Duncan's aides scurrying around collecting this minutiae). The whole pamphlet, written by James S. Duncan, was a public service prepared "so that as broad a section of the Canadian public as possible should have no illusions as to the power of the Soviet drive towards world supremacy."

Duncan was not a political figure and, though given the 131 opportunity, he had deliberately shunned a political career. Yet he found a great satisfaction in appearing on a public platform and in telling his listeners, in fairly condescending terms, what they needed to know. In the case of his Russian trip, the whole ambassadorial exercise took place during a period when his company was in serious trouble. In 1956 profits were to fall to a mere 22 cents a share compared with $1.19 the year before.

In the early days, it would have been difficult to accuse Duncan of negligence, and he was almost certainly not guilty of wilful disregard of Massey's interests now. He continued to have the concerns of the firm at heart, and in his own way directed his efforts towards making a success of Massey-Harris-Ferguson. Nonetheless, it was becoming apparent that Duncan's one-man-rule, his many and varied international involvements, his disinterest in the U.S. operation, and his stubborn refusal to upgrade the quality of his senior staff were all proving to be handicaps to the company's prosperity. Duncan was considered to be indispensable by his staff, by the public, by his peers and, most of all, by himself. But among the people who mattered, the members of the Argus group on the board of directors, opinions of him were changing.

There was in fact a growing antipathy building up between Duncan and all the Argus members. The behavior of the chief of Massey, and his regal ways, seemed to negate the point and purpose of Argus and the investment that had been made. The stake in Massey had been acquired to give Argus control and the ability to exercise its authority, but these objectives were frustrated by Duncan's insistence on running everything his way. Wallace McCutcheon expressed the Argus viewpoint when he told Duncan at one meeting: "What you want, Jimmy, is to have a situation where, if there are three million shares, there are also three million shareholders."

For the time being, these frustrations were kept below the surface. But it was plain that Duncan was creating opposition from some formidable foes; Taylor in particular was not about to subordinate his self-interest and ambitions and defer to Massey's autocratic president. When world farm markets expanded after the war and the company had forged ahead, it had been less easy to criticize Duncan and his management practices. But when markets turned down shortly after the merger with Ferguson, it became apparent that Massey faced a serious economic crunch which would work through to the bottom line and Duncan had no remedies to offer. It was an opportunity for the principal shareholders to go on the offensive. Taylor and Phillips were not slow to express their discontent.

The U.S. and Canadian economies had been expanding

since the war. Suddenly in the mid-1950s they were beset with sky-high inflation and the breakup of a long period of economic growth that coincided with the war in Korea. The situation was particularly grim for agriculture. In the U.S. costs to the farmer were rising rapidly as a result of general levels of inflation. For three successive years, the prices paid to farmers declined. In Canada, the problem of poor prices for grain was compounded by a disastrous crop in 1954. Farm machinery sales were cut back and inventories of unsold equipment began to mount.

Elsewhere, economic conditions were less affected and Massey-Harris-Ferguson, with its international business, was able to continue to draw profits. Nonetheless, the company was in trouble worldwide because of the nature of its merger with Ferguson; few efforts had been made to fuse the companies together and there was no concerted plan to merge manufacturing and distribution into a single efficient operation.* Ferguson's company in the U.S. had been a high-cost, low-profit one. The same was also true of Massey-Harris plants in the U.S. and Canada, many of which were in need of modernizing. Nor was enough attention being paid to costs charged to the company by outside contractors. As a result of the merger, more work than ever was being contracted to suppliers such as Standard Motors in Britain.

In 1954 Massey's profits were nearly $9 million on sales of $300 million, but this overall figure masked a sharp decline in earnings in the U.S. There, profits were down to $700,000 from $2.4 million the year before. If defence contracts were excluded, the U.S. operations were actually in the red by $1.5 million. The trend continued in the following year. Even with defence contracts, there was a loss of $900,000 in 1955, and in 1956 and 1957, as big inventory writeoffs were taken, losses mounted even higher.

For Duncan, this state of affairs was lamentable, but he was convinced that it was a part of the farm cycle, and that tough times had to be accepted along with good times. Certainly, it was no cause for alarm or panic. Farm incomes would recover in due course, and the company's business would improve with them. Meanwhile what was needed was for Massey's management to exhibit faith in the future. Some expansion plans and new projects might have to be curtailed, and costs would have to be held down, but in due course the situation would correct itself.

* Where action was taken, it was often counter-productive. Even before the final papers had been signed on the Ferguson merger, Duncan's North American chief of operations, Herb Bloom, appeared in the Ferguson offices in Detroit and starting handing out dismissal notices. It was a move that did not endear their new Canadian bosses to Ferguson's U.S. staff.

The problem with this argument was that, while it was credible to Duncan and reflected his own experience in the 1920s and 1930s, it did not correspond with what was happening in the rest of the industry where a number of firms were gaining at Massey's expense.

To take one example, profits for Massey from its North American operations were cut in half between 1953 and 1955, while profits of one of its chief rivals, John Deere, were increasing. In 1955, Massey sales were up 10 percent but profits were down 50 percent. For John Deere, sales were up 18 percent and profits were up 37 percent.

The board wanted to see steps taken to improve Massey's financial position. And this was an added cause of strain. The Argus group disagreed sharply with Duncan over dividend policies and the split became almost an annual ritual. On the urging of Taylor and Phillips, one or another directors would make a case for a further increase in dividends. Duncan would counsel that such largesse could not be afforded, and lecture on the need for prudence. As the 1950s opened, he would observe that only three out of fifteen directors had been on the board in the 1930s and were therefore able to recall the parsimony of that time, and the lesson it had taught about building up reserves and maintaining a strong liquidity position.

The board would hear him out, and then promptly vote for an increase in the dividend. By now, Argus members were in firm control of the board, if not the company. They relied heavily on the dividends from Massey, and insisted on them being increased in a regular and reliable fashion. Following the merger with Ferguson, their needs became greater. Argus had arranged for the purchase of Harry Ferguson's interest in Massey and had itself taken another 725,000 shares.

Increasingly, as Massey's difficulties mounted, the Argus directors felt it necessary to intervene. Relations between Duncan and the chairman of the executive committee, Phillips, were generally good, except for the squabble over the dividend policies. Duncan viewed this opposition on dividend payments with resentment, as he did any move that seemed to impinge on his authority as chairman and president, and any statement that implied criticism of him.

In March, 1955, Duncan argued that it was necessary to expand and modernize plants around the world, a move that would make the firm more competitive. He asked for funds for this purpose, and Taylor and Phillips went along with him. A preferred issue was authorized. It sold out quickly and yielded $24 million. The spirit of co-operation and harmony was not, however, to last long. As the year wore on, with no further signs of improvement, the Argus group became more and

134   more alarmed.

At first, their concern centered on the quality of Massey's management, and the lack of any senior executives with business management skills. By the fall of the year, Phillips had forced Duncan to appoint an executive vice-president, and insisted that the incumbent be senior to Herb Bloom. Duncan picked a Ferguson man for the job, Albert Thornbrough. He then listed Thornbrough's duties as being for all practical purposes the same as Bloom's, a typically machiavellian move on his part. Thornbrough was confirmed in his new position by the executive committee.*

The pressure to appoint an executive vice-president was accompanied by entreaties to Duncan to improve the qualifications of his senior staff and replace old-timers such as Bloom and Burgess, which he ignored. He remained studiously loyal to his vice-presidents and proud of them. In Duncan's mind each of them personified the virtues of the self-made man.

In November, 1955, Duncan returned from his grand tour of Russia sick with phlebitis. Over the next six months, his illness was to recur, and it added to the sense of unease both within the firm and on the board of directors. Massey had for years been run as a highly personal undertaking. Without the presence of Duncan, the company drifted.

In the Christmas of 1955, Duncan went off to Jamaica to convalesce. Before going, he instructed Thornbrough to take charge of the budget for 1956 which was to be presented to the board. When Thornbrough began to work on it, he discovered that — with the full knowledge of Duncan — three of the key executives he had to consult with, the general managers of the North American operations, had all gone off on a cruise in the Caribbean with their families. Unable to get sufficient information on how costs could be reduced, Thornbrough was able to draw up only the outline of a budget for the following year, and told Duncan he could do no more.

In the same month, with Duncan absent, Taylor, Phillips, and McCutcheon invited Thornbrough to lunch with them at the Toronto Club to discuss the company's predicament. They also wanted to get his opinions on Duncan. Thornbrough attempted to stay neutral, insisting that he would have to write

* Almost as soon as Thornbrough arrived in Toronto, he was to be made aware of the acrimony between Taylor and Duncan. Returning from a weekend trip out of the city with his family, he was summoned to a private meeting with Taylor who attempted to enlist his aid against Duncan and was sharply critical of him. Notes Thornbrough: "It was obvious that Taylor was manoeuvring to try and get Duncan out, and he kept trying to put words into my mouth. I had to try and walk on eggshells as best I could."

a memorandum to Duncan about the meeting, which he did.
The opinions that he gave, however, did little to calm the
situation. Taylor and Phillips had come to the conclusion that
internal management was in a chaotic state and that Thornbrough,
the only non-Duncan man in the senior ranks, was the best-
qualified and most experienced executive to sort things out. In
these deliberations Duncan saw the makings of a conspiracy
against him. Alternately sick and healthy over the next few
months, he became resentful about being bypassed by the board
and increasingly came to see Thornbrough as a threat to his
position.

A few months later, at a meeting in Detroit, Duncan
confronted Thornbrough in the presence of the rest of his
senior staff. Duncan informed Thornbrough that Massey-Harris
distributors around the world were suffering because the new
MH-30 tractor had not been produced and delivered as promised.
The criticism was an unfair one. Thornbrough was, after all, the
only Ferguson man present and the MH-30 was a Massey-Harris
venture which, through a general lack of initiative by Massey
engineering and manufacturing staff, had not been pushed
forward; even the design work had not been completed. When
Thornbrough pointed this out to Duncan, he made no attempt
to question other executives or put the blame elsewhere. Having
voiced his dissatisfaction with Thornbrough, Duncan simply sat
back and announced that the meeting was over.

By this time, the crisis at Massey was deepening. It had
become apparent that sales targets drawn up by Duncan were
not going to be met, and inventories of unsold equipment were
rising at an alarming rate in the U.S. To add to the company's
troubles, strikes and credit restrictions were reducing business in
the British market. By the middle of 1956, sales were running
below the level of the previous year. Some members of the board
led by Taylor felt that something must be done, and proposed
reducing inventories by cutting prices. There was a great deal
of logic in this since a portion of the excess stocks represented
old or obsolete lines of equipment.

Duncan, however, was bitterly opposed. He regarded the
move as a strategy of defeat. If prices were reduced, it would
be costly and would discredit Massey's name and reputation.
He cited his own experiences in Europe and Argentina during
the thirties, and the recovery of the farm markets there. He
reminded Taylor that he was unfamiliar with the actual operation
of a farm machinery business, while agreeing that — since
Taylor was concerned — he would call a meeting of his senior
sales executives to discuss the matter and report back to the
board. The meeting was duly called. Thornbrough argued that

136 prices should be cut.* All the other vice-presidents supported Duncan.

The result of this stand-off was that Duncan got his way, and Taylor backed down. It was agreed that there would be no across-the-board price reductions. Still, for Duncan, it was a pyrrhic victory. He realized that, on this issue, he was being opposed by a director who was questioning both his ability to run the firm and his competence. For Taylor, the issue was simply one of dollars and cents. Which strategy would work most effectively and yield the best return? But for Duncan it was a blow to his pride and self-esteem, both of which were immense. If the board of directors felt they could run an international farm machinery business better than he could, then let them try.

Forty-six years with Massey.

The finale came in June, 1956. Duncan had been ill and was convalescing at his home. Phillips phoned him and, after inquiring about his health, said that he would like to come and see him two days later. When the two met, Phillips told him that in view of the difficulties the firm faced, the board of directors had decided that the chairmanship and presidency should be split up, and that Duncan's ill health made the holding of both positions too heavy a burden for him to bear. With the agreement of the other directors, Phillips had decided to take on the chairman's job. "Our group is absolutely opposed to the principle of the chairman being the senior executive officer of the company," he told Duncan. "And under these circumstances, we felt that you would not wish to remain on the board."

Duncan replied that he would resign. A few weeks earlier, he had written out the conditions under which he would be prepared to stay on as chairman, president, and chief executive. Stung by Taylor's intervention over pricing policies and convinced that the board was conspiring with Thornbrough behind his back, he had said that in future there must be no questioning of his policies. The Argus group had naturally interpreted this as an ultimatum. And since they were worried about the poor performance of the company, and concerned

* Earlier an incident had taken place that further inflamed Duncan's suspicions about Thornbrough. After an executive committee meeting, Taylor had come to call on Thornbrough and, seeing the door to his office was open, advanced through it announcing: "Al, I want to see you the day after tomorrow for luncheon. I want to go more deeply into this matter of inventories." What Taylor had not seen was that, seated behind the door, was Duncan's administrative assistant, Harry Metcalfe, who immediately reported Taylor's words back to his boss.

as owners and managers, they could not afford to meekly accept it.

Initially, there was no great coalition formed around the idea of removing Duncan or forcing his resignation. The board of directors and the executive committee found it difficult to act. They had very little idea about the operating problems the company faced since Duncan had always believed such matters were his affair and not theirs. Nor could they visualize how the firm could be run without the all-encompassing presence of Duncan. While the Argus men had money and financial connections, none of them — with the exception of Phillips — had great managerial talent.

Duncan's repeated illnesses, however, and the state of paralysis and disorganization within the company forced them to take some kind of initiative. The problem was that any display of initiative on their part would lead to Duncan's resignation. The dispute over pricing policies had already brought relations to the brink. From that point onwards, it was inevitable that any further clashes would cause Duncan to quit.

After seeing Phillips, Duncan called in the company secretary, Norm Appleton, and the assistant director of public relations, Rik Kettle, and told them that his resignation would be announced publicly at the end of the week. No one else was told. The only other people to know were members of the board of directors and the executive vice-president, Albert Thornbrough. The resignation was to be presented to the public and the staff of the company as being caused by Duncan's poor health. No mention was to be made of the disputes and disagreements that had arisen.

On Friday of the same week, Duncan still had a fever when he arrived at the King Street office. Calling all the vice-presidents in for a meeting in his corner office, he made a short speech thanking them for their loyalty to him and telling them that Phillips would be taking over as chairman and chief executive officer.* The vice-presidents were stunned. Though they had known of the rumblings of discontent and the clashes with the board of directors, none of them had suspected that Duncan might go.

Duncan then presided over his last meeting of the board. The details of his resignation and a generous separation agree-

---

* One vice-president whom Duncan did not commend for his loyalty was Thornbrough. Duncan came to see him privately in his office and told him: "Mr. Thornbrough, I hope you are satisfied. You have co-operated with these people [Argus] and now you have got me out." Thornbrough protested that he had not been disloyal or a party to any conspiracy. Duncan retorted: "Weren't you with them all last weekend?" Thornbrough said he was not. On the previous weekend, he had been in Racine, Wisconsin, and not in Toronto. Duncan said he hoped this was so and strode out of the room.

138　ment were drawn up and signed. Then, after forty-six years with Massey, Duncan walked out of the office and down the steps into King Street for the last time. The Duncan era had run its course.*

On the same day, Phillips moved in to the corner office and, in his tough no-nonsense way, began to get down to business.

First, he called in Herb Bloom and fired him. Then, with the other vice-presidents aware of what was happening, he called in Ed Burgess. After a long talk, Burgess came out smiling and told the others: "I've just been fired in the nicest possible way." For the time being, Colonel Phillips' casualty list was kept at two. But, within a year, nearly all of the incumbent vice-presidents had resigned or been removed.

---

* Within a few months of his departure from Massey, Duncan had another job. At the age of sixty-three, the man who had been too sick to preside over Massey in June was by November in command of Ontario Hydro, having been persuaded to be its chairman by the premier of Ontario, Leslie Frost.

9
A Bucket of Worms

Under Phillips and Thornbrough, Massey's objectives were to be as clear-cut and as well-planned as its organization, and in the process the company was to establish itself as a unique corporate hybrid, a Canadian-based internationally-minded global company.

When James Duncan resigned from the company he had been associated with for forty-six years, his career and lifestyle did not change greatly. He may have suffered a personal defeat but his public reputation had not been besmirched and, within a matter of months, he was back at the top of the tree, in command of Ontario Hydro. In his new job Duncan was able to display his talents (his four-year term at Hydro was highlighted by the opening of the St. Lawrence seaway project), and continue to move in high social and political circles with his customary ease and decorum. When he left Hydro, he became an inveterate traveller and writer, eventually retiring to a luxurious home in Bermuda.

During his time at Hydro, Duncan was a masterful publicist for himself. His departure from Massey had rankled him, and he was determined to justify the actions he had taken while chief executive there. As Duncan saw it, the dispute that had arisen at Massey was between the greed and self-interest of the Argus group — in particular, Taylor and Phillips — with their insistence on increasing dividends year after year, and his own wise and restraining counsel. His departure from the company had come at a time when things were improving, as he had predicted they would. His adversaries had taken advantage of a minor break in the company's profitable record, and his own brief period of ill health, to get rid of him.

This interpretation of the state of affairs at Massey did not correspond to the dire problems the company faced. In retrospect, it is perhaps hard to understand why Taylor and Phillips did not move sooner. By the time Duncan was ousted and Massey's difficulties were seen in their proper perspective it was almost too late. In the summer of 1956, Massey was once more close to the brink, and again its very ability to survive was questionable.

The crisis had not been caused by a slump. The end of the Korean War had signalled a last hurrah for the boom times that had persisted for almost a decade. In the farm equipment industry, as in other manufacturing industries, sales had turned down. But none of this could explain the plight Massey was in. Aided by the merger with Ferguson, the company had been able to push its sales forward, but it had been unable, because of incompetent management and numerous inefficiencies, to make a reasonable profit.

James Duncan had been the architect of a great expansion. He had insisted on Massey remaining in the U.S. market during the 1930s; had restored the company's fortunes in Europe in the late 1940s; and had taken advantage of the friendship of Harry Ferguson in the 1950s. To expand, and to go forward, fitted with Duncan's ambitions and inclinations.

But he was very far from being a professional manager, or even a competent one, by the more exacting standards that prevailed in the business world on the 1950s. His reaction to a disappointing performance, and the evidence of a slump in profitability, was to repeat old nostrums about the cyclical nature of the industry and the need to be patient and wait for an upswing in sales. Had he not been proven right in the 1930s? And again after the war? He had always had faith in Massey and believed in the future of the company even when others had not.

Against this kind of intransigence, Taylor and Phillips could only wring their hands. They were concerned about the way things were going and determined to press their own interests. And by the summer of 1956, they had ample reason for their concern and their doubts about Duncan.

In the financial year that had just ended, some farm machinery makers had got themselves into trouble, but none of them faced the squeeze on cash flow and earnings that Massey did. The profit ratio had dropped to a mere 3.3 percent compared with an average 10 percent in the postwar years. There was a chronic shortage of money because sales, particularly in the U.S. lagged far behind the build-up of inventories. The company had to meet the expense of producing equipment it could not sell. A year's supply of Massey-Harris tractors, most of them obsolete, were sitting on dealers' lots or parked in fields outside the U.S.

plants. In the six months before Duncan's resignation, inventories increased by $53 million to $182 million, an astonishingly high level, especially for a company with sales of under $400 million, and this undermined Massey's financial position. If the build-up of inventories continued at this pace, the outcome was clear: Massey-Harris-Ferguson faced bankruptcy.

Duncan's solution was to sit it out and wait for things to improve. But this was plainly untenable. The cash squeeze and the company's exposed position were far too serious. Sales could not be expected to recover quickly enough, or substantially enough. What was clearly needed was to retrench, to eliminate overstaffing and inefficiency, to halt the duplication of effort and expense involved in producing and marketing both the Massey-Harris and Ferguson lines of equipment, and to develop a policy for disposing of unwanted inventories. In the short-term the company had to be brought back to a liquid position. After this had been accomplished, Massey had to be reorganized and given a management structure. It could no longer be run by a single autocratic leader.

With Duncan gone, the task of putting this plan into action now fell to Eric Phillips and Albert Thornbrough. Phillips took the position of chairman and chief executive officer. His business colleague (some would say his business rival since the two began to agree on things far less often), E.P. Taylor, had become chairman of the executive committee of the board. Phillip's role as it developed was something less than that of a chief executive officer since he handed over all responsibility for operations to Thornbrough; nonetheless, the extent of his involvement in Massey did represent a new departure for Argus.

The colonel would come to work in the Massey offices on King Street nearly every day. He would arrive at 9:30 A.M. and stay until lunchtime when he would depart to his private office on St. Clair Avenue to run his business affairs. He plainly relished the job he had assigned to himself and became fascinated by the farm machinery business; he also saw it as an outlet for his talents that would keep him apart from Taylor and enable him to make a mark for himself. As the partnership between the two had developed, Phillips had become increasingly irritated not about Taylor's ability to do deals, but about his inability to follow up or to engage himself in any venture for very long.

When a colleague remarked that Taylor was "the kind of man who starts leaving a room before he's arrived," the colonel appreciated the joke. As an administrator and engineer, Phillips preferred to set himself different challenges, and to work towards more lasting goals. Among the Argus insiders — Taylor, McDougald, McCutcheon, and himself — Phillips was the only one with broader business interests and with experience in manu-

142 facturing and in the management of an industrial enterprise. Alone among the four of them, he was in a position to make good on one of the mandates that the Argus founders had given themselves, that, if the situation required it, they would step in and oversee the administration of the companies in which they had a controlling position. Although such an intervention would take place, nominally, in the interests of all shareholders, in fact, Argus would be safeguarding its own position.

Hard and ruthless when he had to be, Phillips also had an ingrained love of culture and the finer things of life; he was an avid reader, and a highly intelligent debater, and he was a personal friend of painters like Annigoni, Picasso, and Augustus John. At home, he had a collection of Gainsboroughs; a friend recalls visiting him on Saturday morning to find his artist son, Tim, copying a Gainsborough worth $500,000 with the painting propped up on an old bench in the garage. Phillips' other interests included horticulture and agriculture, and he read voraciously on both subjects because he had a farm. He was also a self-made man in the mould of his colleagues at Argus. Starting with the small picture framing business he had inherited from his father, he had moved into glass, and then into supplying safety glass to the automotive industry. It was the sale of his firm to Pittsburgh Glass in the 1940s that had made him wealthy, and given him the capital to join E.P. Taylor.

When the colonel stepped in to take charge of Massey, he was sixty-three years old. Primarily concerned that tough decisions be made, Phillips did not want to get involved in running the company on a day-to-day basis. For this, he turned to Albert Thornbrough, the ex-Ferguson executive who had been isolated and opposed by his colleagues and by Duncan. Thornbrough had hung on at Massey, aware that the situation was moving towards a climax and that Duncan's stonewalling tactics would bring about his downfall. Phillips and Thornbrough scarcely knew each other, but in the aftermath of Duncan's resignation, Thornbrough was confirmed as executive vice-president, and by December, 1956 — when he had gained the confidence of Phillips — he was made president.

Phillips told Thornbrough that the Argus group had a lot of money riding on Massey, and the four of them were determined they would never again subordinate their interests as they had with Duncan. As chief executive officer, Phillips would retain authority over people and management. Phillips took an interest in the principles of organization that were being tried out within the firm for the first time, and in the delegation of authority that was involved; since his days in the British Army in the First World War, he had been proud of his abilities as an organizer and leader of men. And he was involved personally in

operating responsibility for the company to Thornbrough and
the senior managers he set out to recruit.

Drastic changes were needed. Thornbrough, competent
and workmanlike, with a solid background in the industry,
seemed an ideal choice. A low-key and thoughtful midwesterner
he had, in passing through the Ferguson company, become
used to dealing with emergencies and building up a business
organization from scratch. Certainly he had a different
background and personal lifestyle than James Duncan.

Brought up in Kansas during the Depression, he had
graduated from Kansas State College with top honors in
agricultural economics and won a scholarship to Harvard to
study economics. When he arrived in Cambridge, Massachusetts,
in the late 1930s, his sole possession apart from his personal
belongings and clothing was a return rail ticket to Kansas. To
maintain himself at the university — where David Rockefeller
was a classmate and Kenneth Galbraith a young teaching
assistant — Al Thornbrough worked on research projects,
which paid 85 cents an hour, for sixty hours a month instead of
the twenty hours that were supposed to be the maximum. At the
end of his first year, he managed to get the second highest grades
in his class; Paul Samuelson (later to be a renowned economist)
was ahead of him.

Thornbrough's interest in the farm industry, and the close
links that existed between Harvard and Washington, took him
to the U.S. capital in the immediate prewar period, where he
worked with Kenneth Galbraith, who had become deputy
administrator of the Office of Price Administration.*

In the summer of 1941, Thornbrough was given the job of
helping to organize a machinery division, and working on the
pricing of farm goods, which brought him into contact with
leading figures in the industry. During this period, as a young
civil servant, he first met James Duncan in Washington. The
position he held was not one calculated to endear him to the
farm equipment makers who had no great liking for
Washington's price controllers. Nonetheless, his work impressed
several of the big firms. When war was declared, Thornbrough
went into the army and rose to the rank of lieutenant-colonel
in the Corps of Engineers. When he left the army in 1946,
he was offered several jobs by farm companies including International Harvester and Deere.

---

* At Harvard, Thornbrough had set out to do his dissertation on the economics of
technological change in farming, using his background as an agricultural economist.
Galbraith knew of his plan and called him from Washington. In his new position
as a price controller, Galbraith wanted to draw on Thornbrough's work and use it
to confront the big farm machinery companies.

The job offer that he eventually accepted showed him to be a young man keen for adventure. He joined Harry Ferguson's U.S. company, becoming an assistant to its U.S. chief, Roger Kyes. He then immediately became embroiled in the break-up between Ferguson and Ford. Kyes was an outstanding personality, and an articulate and driving salesman. He ran Ferguson's Detroit operation with firmness and ruthlessness (he was known to employees as "Jolly Roger, the hatchet man"), and he pushed sales relentlessly, creating an acceptance of the Ferguson System in the U.S. where none had existed before. Kyes' qualities as a salesman and manager were subsequently to take him to General Motors,* and to Washington as deputy secretary of defence. It was after Kyes' departure from the Ferguson presidency in Detroit that Harry Ferguson had first tried to interest Duncan in joining the company.

For Thornbrough, working under Kyes was a rewarding experience. He was given a great deal of responsibility, and he was pitched into a situation fraught with trouble and confusion. Harry Ferguson Incorporated was desperately undercapitalized and was constantly being thrown off balance by its precarious position in the U.S., and by the whims and eccentricities of its founder. Thornbrough was given the job of finding an American gasoline engine to put in the Ferguson tractor; outside the U.S., nearly all tractors were diesel-powered so Ferguson's British contractors could not supply one. Thornbrough got the engine from Continental Motors. He was also given the job of procuring components for Ferguson after the break with Ford. It was a task that required a great deal of diplomacy. Many suppliers were frightened that if they sold Ferguson parts they would be cut off from doing any business with Ford. Thornbrough patiently negotiated for transmissions, rear axles, hydraulic pump assemblies, and dozens of other essential components. It was due to his efforts that, within a year of severing all ties with Ford, Harry Ferguson Incorporated was back in business. The first wholly Ferguson-assembled tractor came off the production line in October, 1948.

Thornbrough had made his mark with Ferguson. And he had gained invaluable experience. As an executive, he may have lacked the cosmopolitan background and the polish that

---

* At General Motors, Kyes ran up against the auto industry's flamboyant John Z. de Lorean who was later to quit and start up his own firm in Northern Ireland making luxury sportscars. In J. Patrick Wright's book *On a Clear Day You Can See General Motors* (New York: Avon Books, 1980), de Lorean's experience with Kyes is recounted: "He [Roger Kyes] made life unbearable for me, and he was dedicated to getting me fired; he told me so, many times. Fortunately I had the protection of my ability...to fend off Kyes. But I remember vividly my conflicts with him, especially when he was irritated by my style of dress. The corporate rule was dark suits, light shirts, and muted ties. I followed the rule to the letter, only I wore stylish Italian-cut suits, wide-collared off-white shirts and wide ties. 'Goddamnit, John,' he'd yell, 'Can't you dress like a businessman? And get your hair cut too.' "

seemed to be required to run a major international firm;
however, in the parlous state Massey found itself in during
1956, Albert Thornbrough did have other more important
capabilities. Skilled in implementing new approaches and new
systems, he was adept at creating order out of chaos.

Over the next three years, Eric Phillips and Albert
Thornbrough were to transform Massey into an efficient and
well-managed firm, and they were to prepare the way for a new
flowering of Massey's overseas operations. The critical year
was 1959. By then, most of the work had been done in
Toronto and in the U.S., and the company was to begin
expanding its international business on a great scale then ever.
But first a lot of hard work was needed to solve Massey's
immediate problems, and bring the company back from the
brink once again.

Enter Massey-Ferguson.

Duncan's claim that he had been driven out of Massey
needlessly, and that he had left the company in good shape,
never failed to irritate Eric Phillips. He was an astute and
powerful man who did not believe in going public on such
matters, or engaging in senseless slanging matches. Duncan's
actions made him bitter and led to the ending of their friendship.
Some years later, when E.P. Neufeld was commissioned to write
a book on Massey (*A Global Corporation*), Phillips tried to stop
him from even talking to Duncan. The man who had led the
company for nearly two decades had become, like Stalin in the
Kremlin, a non-person. Neufeld did eventually get to talk to
Duncan, but it was only because Thornbrough intervened and
insisted that he must.

Phillips' ire was understandable. When he moved into the
King Street office and began reviewing the measures needed
to save Massey, the company's difficulties seemed very apparent.
After firing Duncan's chief lieutenants, the first task was to
restore the company's financial position, and this had to be done
quickly. Phillips therefore gave an order to Thornbrough to
do anything that needed to be done to make Massey solvent.

Basically, there were two ways of cutting Massey's losses;
either sales had to be increased substantially, or plants would
have to be closed down. The previous management had worked
on the belief that sales would stage a recovery, but the prospect
for a climb-back that would make much difference and lead to a
reduction of inventories was remote. The second option, shutting

down plants, was the only viable course to take. Thornbrough ordered the closing of the plants in Racine, Detroit, and Toronto. And he decided that Massey's second-largest plant in the U.S. the high-cost Batavia manufacturing operation, would have to be shut down permanently.

As production came to a halt, the build-up in inventories ceased, easing the financial pressures. Thornbrough then proceeded to mount a bargain-sale campaign, slashing prices and reducing inventories of old and surplus equipment. In Detroit, there were 5,000 tractors in a field outside the Ferguson plant, all of them with faulty parts and rusted by the weather. They were repaired and cleaned up and sold off at firesale prices. To get rid of its huge inventories, the company had to pay a price. Writedowns of $6.5 million were taken in 1956 and $13.6 million in 1957. But the equipment was moving out of the factories and off the dealers' lots, and cash was flowing back to Toronto. In the four months between June and October of 1956, the company's net cash position in the U.S. and Canada improved by $40 million, and its total outstanding bank loans came down by $50 million.

The sell-off eased the cash squeeze and solved the immediate crisis. It also gave a considerable fillip to the new management since it was the course favored by Taylor and Thornbrough a year earlier, and dismissed by Duncan as impractical. There were, however, greater problems with the North American operations, and these had to be tackled if Massey was to operated profitably again.

Figures for 1957 showed just how drastic the North American situation was. While the company was doing well elsewhere, its losses in North America including writedowns for getting rid of inventory and the reorganization of its distribution system were a horrendous $15.7 million. Only profits from aboard and a $5 million tax credit enabled Massey to escape with a net deficit no bigger than $4.7 million. Ordinary operating losses in the U.S. and Canada, by themselves, amounted to $4.8 million.

This disastrous performance had taken place in a year not especially poor for the farm industry. A writer for *Fortune* magazine, William B. Harris, noted in an article written the following year: "The simple and sad fact was that Massey's American and Canadian operations had been so badly managed for such a long period of time that, when the farm-machinery business came down off its high level after Korea, the firm was totally unprepared for the period of tough competition that ensued." Massey was indeed uncompetitive. Just how uncompetitive, Thornbrough was rapidly finding out.

One of the first things he discovered was that Toronto

headquarters did not know, and had no way of finding out, how large inventories were and how rapidly they were piling up. There were no records at all. Management in Toronto had to rely on frantic messages from branch offices. Only when these were received did they realize the company was in trouble. This lack of control extended to the factories. There was no system of cost control, and the company did not know whether it was manufacturing goods in any particular plant at a profit or a loss.

The worst situation of all, from a competitive standpoint, was the dual distribution system inherited after the merger with Ferguson, which the previous management had sought to keep. The reasoning behind its retention was that, like a car company, Massey-Harris-Ferguson could continue to offer different lines of equipment and a choice of different models to farmers. In practice, the red machines of Massey and the grey machines of Ferguson were fighting each other, with duplicate organizations from manufacturing to sales. Since their combined strength was not as big as International Harvester, or Ford, or John Deere, it was an expensive and inefficient way to operate.

Thornbrough established inventory and cost control systems. He raised pay scales at the executive level and on the factory floor, putting in new bonus and incentive plans. And he integrated production which had been split up among a number of plants. Then, in early 1957, he turned his attention to ending the dual distribution system and creating a single company.

Time was of the essence. The decision had been made to establish the company name as Massey-Ferguson, thus ending the cumbersome Massey-Harris-Ferguson and, in the fall of 1957, a new tractor and implement line were to be launched and marketed under the streamlined name. This meant that within six months a single distribution and dealer network had to be forged, and the old arrangements with the Ferguson distributors phased out.

Thornbrough had hired a Texan, John Shiner, from Ford to take charge of corporate marketing. He had also won agreement from the board that the Ferguson distributors had to be bought out despite the expense. In March, Thornbrough and Shiner set off in the company's Lodestar and journeyed to the Midwest to persuade, cajole, and sometimes threaten the independent businessmen who were franchised distributors for Ferguson to sell out. In all, there were seventeen Ferguson distributors. Between them, they employed 500 people, and had over 1,800 dealers. And virtually none of them wanted to sell.

The talks went on for days. Thornbrough had to promise

148 to look after old employees, to take over and sell buildings, to buy out leases, to pay money for inventories of machinery and parts, and he had to give some Ferguson dealers the chance to switch to Massey-Ferguson. The negotiations were tricky and Thornbrough and Shiner had to proceed carefully. They were concerned that the distributors might complain to the U.S. Federal Trade Commission. If so, Massey could have been accused of restraining trade and of monopolistic practices.*

When it was all over, the pair had travelled 17,000 miles. They had bought out all but one of the distributors for just over $2 million more than they were worth. The remaining holdout, Southland Tractors of Memphis, insisted on suing Massey on an anti-trust charge. The Southland suit was eventually settled out of court for $250,000.

The result of these manoeuvres was to transform Massey into a lean and hungry sales organization. Compared with its previous oversize sales operation, the difference was astounding. From fifty-three main sales offices in the U.S. and Canada, the number was thinned down to thirty-four branch offices and eight sub-branches; from 5,700 dealers the number fell to 3,800; and from 2,700 employees the number dropped to 1,600. Despite these reductions, sales in the U.S. rose by 30 percent in 1958, and in Canada by 9 percent.

Greater efficiency in both Massey's manufacturing plants and its sales organization had been the responsibility of Thornbrough, but he had very quickly brought in experienced executives from outside, most of them Americans, and he had planned several key moves with the aid of management consultants from McKinsey and Company. Modern management techniques were being applied to running Massey for the first time, and it was a novel experience. To those who watched the company, and were interested in its unique position in Canada and its international standing, it seemed that Massey-Ferguson was being Americanized. But few could argue with the results.

Eric Phillips and his Argus colleagues were certainly not about to complain. Between 1957 and 1959, the company went from a loss of $4.7 million on sales of $412 million to a profit of $21 million on sales of $491 million. The recovery had been

* The legal complications in the U.S. stemmed from the fact that Ferguson had operated through distributors who were independent businessmen and had their own dealer network, while Massey-Harris had a wholesale branch system in which dealers were franchised. U.S. law held that the firm ultimately controlling a product had to sell it at the same price to the same category of customer; with its differing distribution and discount structure Massey-Harris-Ferguson was technically in violation of this following the merger. Aware that the company was trying to get itself out of a delicate situation, some Ferguson distributors resorted to rough tactics to press their case and raise the ante. In Florida, one distributor bugged a private meeting with Thornbrough and Shiner and threatened to use the tape in court. As it turned out, the tape was garbled and the firm decided to make a settlement instead.

achieved.   149

Phillips himself had done more than merely preside over things and approve the initiatives being put forward by others; he had himself drawn up the order of battle. Previously, Massey had never had an organization chart of any kind. In the past, all decisions of importance had rested with Duncan, and no attempt had been made to adopt the standard management practice of dividing staff and line functions. On July 12, 1956, at the urging of Phillips, the first organization chart was drawn up. Before the end of the month a central co-ordinating committee had been established to advise Thornbrough on company policy and overall objectives.

Phillips was a strong believer in delegating authority and in decentralization. But, in the wake of the Duncan years, he recognized that the company lacked experienced senior managers. In a memo to Thornbrough he noted: "What we urgently need are a few competent people with some talent for the executive functions."

This dearth of skilled senior executives was to be remedied as one by one the Duncan managers either left or were forced out, to be replaced by a new breed of business-school-educated managers, nearly all of them Americans. Many of them came out of the auto industry since production methods and merchandising were similar, but others were drawn from investment and from fields such as advertising.

For the new arrivals there was the challenge of being able to reshape a company where the slate had been wiped clean, where management concepts were being introduced for the first time, and where there were plainly opportunities for improvement and for advancement as the company modernized its operations and expanded.* In December of 1956, Phillips and Thornbrough circulated a Memorandum of Organization that was to be regarded as the Bible on company structure and the delegation of authority.

The memorandum clarified the hazy demarcation that had existed between Toronto and the various overseas operating units. Henceforth marketing and manufacturing activities, together with some support services, were to be organized locally in response to the needs of the market. Any activities that would shape the long-range character of the company, such as control over product line, over facilities and money, and the planning of strategies to fit new opportunities internationally, were to be centralized.

Out of this re-evaluation, and the management personnel

---

* One executive who joined the company as Thornbrough's assistant and rose to be its senior vice-president, John Staiger, says of this period that is was highly creative for all the newcomers: "What was wonderful was that we could do anything and it would be an improvement on what had gone before."

that came in, a new philosophy and direction began to emerge. As the immediate worries lessened, Phillips and Thornbrough found time to draw up basic plans of how Massey would move into the 1960s, and how the company would structure itself to handle its major competitors.

The main lines of this approach were formed as early as 1957, only a year after Duncan's departure, and they were to change very little subsequently. For the next twenty years, Massey-Ferguson was to pursue a course that corresponded to these principles, and it was to carry on in pursuit of expanding these goals even when they led to the brink of disaster in the late 1970s. The basic goals were first, that product planning and development should be adapted to local market needs but integrated worldwide through head office control; second, that a worldwide network of factories should be established, with production integrated and rationalized so as to maximize profits under varying economic and political conditions; and third, that there should be a worldwide structure of manufacturing and marketing units, run by nationals, in all major markets.

While there was nothing wrong with these objectives in themselves, they tended to sanction growth for its own sake, to legitimize the notion that Massey could operate profitably in any country at any time because production could be integrated worldwide. In fact such economies of scale were to be very hard to achieve. And the larger Massey became the more vulnerable it was to economic and political conditions on a global scale.

Hire and fire and generally raise hell.

During the winter months of 1957, Massey was in a state of upheaval. As the new team of Phillips and Thornbrough concentrated on reorganizing their loss-making U.S. and Canadian operations, many senior managers with long years of experience were forced to go or voluntarily relinquished their jobs. It was the biggest exodus that had ever taken place at Massey; staff at head office changed and the separate Canadian and U.S. divisions were combined into one North American division, forcing many old-guard managers out simply by abolishing their jobs. The turmoil drew comments in the press. "More Shake-ups at Massey-Harris" announced a headline in the *Toronto Star* on February 9, 1957.

Within the firm, there was a great deal of dissension, and

quite naturally, worry and concern. Often it was difficult for 151
senior executives to understand or interpret what was happening,
or appreciate why change was necessary. The old virtues at
Massey had always been those of self-help and salesmanship;
suddenly a new language had sprung up which put as much
emphasis on decentralization, delegation of authority, and
overhead cost controls. The new approach was bewildering to
many company men, and not just to those closest to the scene of
the action in head office. Plaintive memos were received from
the field, from "old Massey-Harris" men and "old Ferguson"
men wondering what was to become of them under the new
regime.

Inevitably, old rivalries began to surface. Since there
had been no consolidation of the reds and the greys, Massey
and Ferguson employees did not think of themselves as part
of a unified company. It was easy for discord to take root within
the ranks of senior executives. In the early days, the feeling
had been that Massey had won out over Ferguson and that
Massey men had acquired most of the top positions. This state
of affairs was attributed, with some venom, by the Fergusonites
to the "old man" (Harry Ferguson) who had "sold them down
the river." When Thornbrough took over, he brought in two of
his Ferguson colleagues to fill the topmost jobs: Charlie
Herrmeyer became vice-president of finance and Hermann
Klemm became vice-president of engineering. Suddenly the
situation was reversed; now the Massey men saw their influence
waning.

After the amalgamation and the emergence of the company
as Massey-Ferguson, the rivalry died down. Thornbrough himself did his best to dispel it. Generally, the company began to
operated in all its markets as a single unit, with an identifiable
organization. By this time, however, a rapid transformation
had taken place. Neither the former Massey executives, nor the
former Ferguson executives seemed to dominate anymore;
instead the positions of power were held by new men, many
of whom had never worked in the farm machinery business but
had acquired their experience in U.S. multinational companies
like Ford, and American Motors, and Goodyear.

The break with the past was clear and distinct. The tradition
in Massey had been for senior posts in the company to go
to people with a background in the farm industry, or because
of loyalties and links they had established with Duncan. The
company had almost never looked outside for management
talent (The only senior outsider had been Guy Bevan who,
after many years with the company, was still regarded as an
interloper). Thornbrough advised Phillips that certain appointments would go to Canadians, in public relations and personnel,

152 and he would try to recruit others. But the bulk of management talent in areas such as marketing and manufacturing was in the U.S., and for a company of Massey's size and potential, it would be necessary to look there for executive material.* After the firings, and the departure of so many old-guard executives, the hiring of the senior staff had to be done quickly.

Phillips' reply to this was that he did not care where the new managers came from, or their nationality. As a multi-national company Massey would open its doors to Americans or Europeans as much as Canadians.

The hiring policy that began to take shape emphasized not only professional managers but also those experienced in solving problems and in reconstructing companies that had run into trouble. If there was any single industry which met Massey's recruitment needs most it was the auto industry. In terms of management styles, there was a great deal of similarity between the farm equipment and auto businesses. Moreover, the car companies were ruthless where staff was concerned; problems in the industry were solved by installing new managers with full authority to hire and fire and generally raise hell. Massey needed a new management corps and it need some hell-raising. If it was to survive and grow, the company had to do things differently. The days of relying on tradition and on staff and customer loyalties were over.

On the engineering side, there was some stability. Hermann Klemm, a bright German-born aeronautical engineer, took over in North America as vice-president of engineering; he had been the chief engineer for Harry Ferguson Incorporated in Detroit and had been responsible for putting the popular FE 35 tractor into production the previous year. His promotion led to the dispatch of Guy Bevan to Europe to head engineering operations there. The move was to have important repercussion later since the highly social Bevan, a former Cambridge University oarsman, knew many prominent British industrialists personally.

Thornbrough also brought Charlie Herrmeyer with him as vice-president of finance. They had been friends as well as colleagues at Ferguson, but the relationship between them was to become more difficult. Charlie Herrmeyer was a perfectionist, a man of rigid views about his own job, and about the primacy of the vice-president of finance in the financial area

---

* In an effort to get changes made, Thornbrough had approached Duncan ten days before his resignation and told him that the company should find a vice-president of manufacturing from outside. Ed Burgess, Duncan's wartime colleague, nominally held the post but he was not, in Thornbrough's opinion, a professional manufacturing man of the calibre that was needed. Duncan, who was recuperating from his illness at home, told him: "Mr. Thornbrough, this is absolutely the worst thing you can do. If you persist and do this, you will lose the loyalty of every single member of the organization." To which Thornbrough replied: "The job has to be done, Mr. Duncan, and loyalty alone will not do it."

of the company. Thornbrough agreed that his views should be
listened to, but he reserved the right to overrule him. There
were disagreements about a planned move into light industrial
equipment and about new planning procedures, disagreements
that finally brought about Herrmeyer's resignation. Herrmeyer
was replaced by another experienced financial man, Ken Tiffany.
Again the relationship that developed was an uneasy one.
Thornbrough was keen on expansion, and was carried away
by the logistics of moving into new markets, and the progress
that could be achieved in new countries.

Thornbrough's vice-presidents of finance counselled
caution. They were worried about the company overreaching
itself, and about where the money was going to come from.
Thornbrough pressed his case, and nearly always got his
way. But his relationship with his chief financial officers was
never comfortable, a fact that was to have great significance
for Massey later on.

Aside from Klemm and Herrmeyer, the key positions were
all filled by newcomers, a number of whom came in during
early 1957. John Shiner had joined in December of the previous
year to extend the sales organization into a fully fledged
marketing one. Shiner was a robust and gregarious Texan who
had been a vice-president in Ford's Mercury Lincoln Division in
Chicago. The first outsider to occupy a senior position at
Massey for many years, Shiner was a brilliant salesman. He had
got to know Thornbrough several years prior to his appointment
at Ford when he was working for Goodyear in Detroit. Shiner was
familiar with the production problems involved in a large manufacturing and assembly operation, and he knew what needed to be
done to market and merchandise expensive machinery. Gone
were the old days of simply presenting a product for a salesman
to dispose of. The sales effort had to be backed up with market
research, with forecasts and planned product changes, with
better service and credit facilities, farm management counselling,
and an improved distribution network.

Suddenly, Massey had moved into mass merchandising. It
was not enough to produce a piece of farm equipment and
sell it. Greater attention had to be given to the customer and
his needs, and this had to shape the product line and
company's engineering strategy. Another American, Ed Barger,
a former professor of agricultural engineering, was put in
charge of product planning. As well, Massey had to promote
itself and, since it was in competition in the U.S. with some
giant corporations, it had to stand out as a recognizable and
different company. It could not just rely, as it had done since
the days of Hart Massey, on the belief that its products were
superior and would sell themselves.

In marketing, an element of fun and showmanship began to creep in. Ward Dworshak, the son of an Idaho senator, was brought from Ford's farm machinery division to run U.S. sales (the redoubtable "Marshal Joe" Tucker had retired). Dworshak ordered all salesmen to wear ties, together with matching coats and pants. Since at the same time, salaries were raised and salesmen were given the use of a leased car, the new style of dress could be afforded. This move raised Massey's image considerably. Within a few months, Dworshak was swamped with applications for selling jobs from college graduates and experienced sales representatives. Massey was an organization that offered money and sartorial style.

Another recruit was an advertising man, George Gordon, who had attended Wharton School of Finance at the University of Pennsylvania and moved from there to a Madison Avenue agency where he had handled the Procter and Gamble soap account. George got to know John Shiner over a game of tennis at a resort hotel, and Shiner persuaded him to come to Toronto and join Massey. Gordon knew nothing about manufacturing, and he had never been near a tractor in his life, but Shiner had the intuition that Gordon, a young man in his early thirties, would shake things up.

And he was right. George Gordon would attend meetings and come out with bizarre ideas for merchandising and promoting Massey. To him, selling tractors was like selling soap, or lingerie, or sports cars. A little razzle-dazzle was needed to perk things up and put the farm machinery industry on its feet. His colleagues reacted with horror, but George pressed on with his proposals. A travelling show of dancing girls toured the U.S. to help sell Massey-Ferguson tractors; a country music television show was sponsored with its star, Red Foley, promising a cheque for $100 to any farmer who bought a MF-35 tractor; dealers and their wives were invited to a show of progress in Detroit, an extravaganza which — according to Massey publicity — involved the biggest one-day civilian airlift ever undertaken. Then there were the lesser events — the bonus banquets, the parade for profits, and the bonus fairs.

George Gordon eventually moved on, becoming a marketing man for an airline instead of a farm machinery maker. Nonetheless, in his few years at Massey he had gingered things along and done wonders for the company's public profile. In the U.S., the company had been the first in the industry to use national television to promote itself. And there were many who argued that it was Red Foley and his $100 cheques that turned the company around in the late 1950s, achieving more than any corporate reorganization could ever do.

Other Americans from outside the industry also moved

into top posts. Hall Wallace became vice-president of manufacturing. A graduate of the Massachusetts Institute of
Technology, Wallace had worked with Allis-Chalmers and
Johnson and Johnson. He was a highly qualified engineer and
a man of great physical toughness, a boxer and footballer, who
commanded intense loyalty from his factory managers. Until
his appointment, Massey factory managers had been low in the
hierarchy and were generally unregarded by management in
Toronto. Wallace was given a free hand to change this.
Managers became responsible for the total operation of their
factories including all purchasing.

The changes were not universally welcomed. Factories had
to be closed down, not only at Batavia in the U.S., but also
in Ontario at Market Street in Brantford and at Woodstock.
The company had committed itself to producing globally
rather than locally, and to integrating production to save
costs and gain efficiency.

New wage rates were introduced to replace the old piecework system of payments, a move that was not popular with
the employees at first. In 1958, the labor unions in Canada took
their grievances to a conciliation board and strikes were
threatened. Still, the new ideas and methods of Wallace, and
another newcomer, John Belford, who had come from Canadian
National Railways to be Massey's director of personnel and
industrial relations, won out.* Like any large company, Massey
was to have its recurring labor disputes and strikes. But at
this crucial period in the late 1950s, the company did succeed
in changing direction and introducing systems of organization
and pay scales that did increase its operating efficiency.

Many of the new administrative procedures were the brain-
child of yet another American executive from the automobile
industry, John Staiger. A bluff, no-nonsense character, Staiger
had graduated from the University of Dubuque, then taken
postgraduate courses in accounting and business administration
at a succession of midwestern universities, Marquette, Wisconsin,
Michigan, and Chicago. Before joining Massey, he had been
assistant to the hard-driving George Romney at American
Motors, and he arrived in Toronto to occupy a similar post as
Thornbrough's assistant dealing with planning and administration.

* To make good on his promise about appointing two Canadians to senior posts,
Thornbrough selected Belford, a McGill graduate, for director of personnel and
industrial relations, and Rik Kettle, the former Upper Canada College schoolmaster
who had been recruited by Duncan, to be director of public relations. Kettle had
been one of only four people (the others were Phillips, Thornbrough, and the company
secretary, Norm Appleton) to know about Duncan's resignation in advance, since he
had the job of releasing the information to the press. At the time, Phillips promoted
Kettle, and told him to draft another press release announcing his appointment.

156   In the early days, Staiger acted as a kind of liaison man. He had known Thornbrough from the days when they had been neighbors in Detroit's Grosse Pointe. Besides being friends, they had also shared a common interest in their jobs; both were working for struggling companies, Thornbrough for Ferguson and Staiger for American Motors. In Staiger's case, his business background had given him an intimate knowledge of how to handle such things as mergers and plant closures, knowledge that could be directly applied to the situation facing Massey.*

After Thornbrough's move to Toronto, the two had kept in contact. When Thornbrough took over the presidency at Massey, Staiger sent a congratulatory letter. Within a few months, Thornbrough was back in touch with him. He wanted his new executives, including Shiner and Wallace, to have the benefit of Staiger's experience, as well as access to some of the manuals and procedures Staiger had developed at American Motors. The Massey executives visited Detroit and Staiger also journeyed to Toronto as an informal and unpaid consultant.

The problems that Massey faced were compounded by the lack of information and controls at head office, where decisions were made against a background of little or no knowledge. In the past, budgets had been drawn up by the senior finance staff who knew virtually nothing about the company's operations or what was happening in the marketplace. Invariably, budgeting was months behind rather than months ahead.

When Staiger finally joined Massey as Thornbrough's assistant, he was to spearhead the imposition of new cost controls and manpower planning, and to provide the company with a parts operation modelled on those of the major auto companies. He was typical of the recruits in that he knew little about Massey (despite an upbringing in rural Iowa), and less about Canada and the Canadian business scene. Staiger was also two years older than Thornbrough. Although he characterized the state of Massey at the time as being "a bucket of worms," Staiger was attracted to the job and to working for Massey. He regarded the creation of a modern efficiently managed firm as a more formidable challenge than anything he would face at American Motors. Even though going to Toronto would mean stepping out of the mainstream of a U.S. business career after more than twenty years in the auto industry, in Staiger's estimation, the challenge was worth it.

Massey's new executive team had been built up quickly. By the time Staiger joined in the spring of 1957, it was

* Staiger's participation in crises at American Motors had been numerous. The company had just staved off a management takeover attempt and he had completed a two-year assignment attempting to turn around the company's troubled automotive division. He had also worked on a major plant closure in Detroit, on the merger with the Hudson Motor Company, and on a contemplated merger with Packard.

virtually complete. The team that Thornbrough gathered together set itself three main priorities: first, to improve the products Massey manufactured; second, to upgrade the manufacturing process itself; and third, to improve the servicing of the product in the field. To accomplish these goals, Massey's new executives realized it would be necessary to force their innovations and systems downwards through the company, instilling new ideas and concepts into people in the field and the factory who were used to doing things their way and gleaning their information haphazardly, from jottings scribbled at random in little black notebooks. Now, under Phillips and Thornbrough, Massey's objectives were to be as clear-cut and as well-planned as its organization, and the strategy for accomplishing these objectives was likewise to be tightly planned and co-ordinated. In the process the company was to establish itself as a unique corporate hybrid, a Canadian-based, internationally-minded global company.

In the 1960s, Harvard's business students would be presented with Massey-Ferguson as a case study of how, from unlikely origins, great multinational companies could grow and flourish. In 1968, when the British-based magazine *Management Today* sought to explain the relevance of Massey, it described the company in the following way: "Massey-Ferguson provides the extraordinary spectacle of American businessmen, exiled in Toronto, taking on all the might of U.S.-based companies in the same line of business — and taking them on in very similar circumstances to the ones in which the larger European companies have to operate." After describing the process involved in bringing about this extraordinary transformation and noting that, "today half of the top management are U.S. citizens," the magazine concluded: "In some ways, Massey-Ferguson shows what the American manager might do if he had to face the kind of challenge represented by America."

The tribute was by no means misplaced since, in the next decade, the company was to enter into a period of extraordinary international growth. In a strange twist, the Americanization of Massey under Thornbrough was to signify the coming of age of the company as a multinational firm, and its further expansion was to be far more impressive under the direction of this onetime farm boy from Kansas than under the peripatetic James Duncan.

Part Three:
The Thornbrough Years (1956-1978)

10

The One True Multinational

The company was suddenly to find itself becoming a major force in manufacturing in another country, Britain, a process that over the next decade was to make it the most important farm machinery company in the world and a direct competition to the two giants of the industry in the U.S.

In the farm machinery industry, there is one competitive reality that is as real to Massey-Ferguson today as it was to the firm in the 1920s and 1930s; the U.S. is the single dominant factor in the global agricultural business. Nationwide its farmers have the greatest, most productive land to work with and, consequently, they derive an income from it that is a magnet for all farm equipment manufacturers.

Some of the farm equipment companies draw a major portion of their revenues from the U.S. farm belt, in particular the rich land stretching from Iowa to Ohio known as the corn belt, while others are constantly struggling to maintain even a marginal share. But in the evolution of the industry there have been only two consistent winners — International Harvester and John Deere.

From the beginning, International Harvester has ranked as the giant of the industry; an amalgam of small companies that combined to give it a powerful and entrenched position, then diversified into making trucks and construction equipment. The company has suffered poor years and financial mishaps. But its worldwide position has not been seriously challenged.

John Deere has reigned supreme in particular localities. It has woven its green and yellow emblem into the political fabric of the state of Illinois, and has come out ahead of International Harvester in farm equipment in the U.S. However,

Deere has had recurring problems whenever it has attempted to
move overseas.

The two companies have fought with each other, but they have never had to worry about being shouldered out by anyone else. This is still the case today as it was in the late 1950s when, having patched up the immediate problems at Massey, Thornbrough and his aides began making plans for the company's future growth. Massey-Ferguson held a dominant position in Canada, but as Hart Massey had recognized in the 1890s, the Canadian market did not provide even a modest base for a global operation. The merger with Ferguson had given Massey additional strength in Europe, and particularly in Britain. Certainly it could capitalize on this. But what form should future expansion take? And how could the company mount an effective challenge in the U.S.? Obviously, with a third of all free world farm machinery sales taking place in the U.S. three-quarters of them being replacement sales in the corn belt, the American market could not be ignored.

The strategy that eventually took shape under Thornbrough called for Massey to move in on the target indirectly and under cover. Massey was not in a position to embark on a head-on contest with John Deere and International Harvester. It did not have the products or the distribution system to match them. Very well, reasoned Thornbrough, it would leave the field of battle and seek reinforcements elsewhere. Then, later on, it would return to attack in force.

Massey had its traditional areas of strength; in Canada, and in its manufacturing complexes in Europe and especially in Britain, as well as a long standing reputation as a supplier and manufacturer in South America, Africa, and Australia. It could therefore concentrate on developing these strengths. In doing so, it would gather together a larger cash flow from its international operations to support a new drive in the U.S.

The company's position had serious weaknesses too, and these did not simply involve its failure to make much impact in the U.S. It had inherited the Ferguson System and an impressive name and distribution network to go with it, but it had also inherited the shortcomings of Harry Ferguson's way of doing business. Ferguson had patented his system but he had never had any desire to go into manufacturing for himself. Instead he had relied on others to supply him with his Ferguson-designed equipment and components. To some extent, Massey-Harris had followed the same path. Despite the strength of the combined company's sales, neither Ferguson nor Massey had any plants for manufacturing engines or gears and transmissions. As a result, both were independent on outside suppliers and contractors. When suppliers experienced labor problems and strikes, both companies suffered. Moreover, there was always concern about the price and quality of the components they were receiving. Production of Massey-Ferguson tractors and other farm equipment was at the mercy of others.

This situation was far from satisfactory and had to be rectified. No great expansion of the company could be contemplated unless it had greater control over outside manufacturing and sources of supply. As a defensive measure, if for no other reason, Massey had to become more self-sufficient. To the managment in Toronto, this was not at first as obvious as it seemed; they were, in any case, busy with internal problems, and with the reorganization of the company. For these reasons, the growth that began to take shape in the late 1950s started involuntarily and was not actively planned nor directed.

What it did do, however, was to launch Massey on the drive towards its eventual goal; that of being a successful international firm able to challenge its larger rivals in the U.S. corn belt. Acting defensively at first, the company was to find itself on the road to becoming a major force in manufacturing in Britain, a process that over the next decade was to make Massey the most important farm machinery company in the world, and a direct competitor to the two industry giants in the U.S.

The struggle for Standard.

When Massey-Harris merged with Ferguson in 1953, the British newspapers had been told it was the deal of the century for the farm machinery industry. The exaggeration was pardonable under the circumstances. Among Britain's glamor industries, the manufacture and distribution of tractors did not rank very high. Although Harry Ferguson was a colorful, newsworthy figure, the public relations men, and the participants themselves, had still found it necessary to embellish the story of the merging of the two farm machinery empires. For the merger to get public attention, it had to be classified as an epoch-making event.

In 1959, Massey-Ferguson began to involve itself in something which, to the British way of thinking, was much more controversial and sensational — the automobile industry. The involvement came about because of Harry Ferguson. Once Massey had committed itself to buying Ferguson's business and then buying him out, it also had to cope with the web of interrelationships that had grown up around the business, the most important of which was with Standard-Triumph, an

automobile company with a cherished place in Britain's
corporate firmament.

The Standard Motor Company was founded in Coventry in 1903, ran into disastrous times during the Depression when it was rescued by a young general manager called John (later Sir John) Black, and after the war took over the assets of the bankrupt Triumph Motor Company. Standard-Triumph had a problem which it shared with other car makers in the fragmented British industry, among them the Rootes Group and the Rover Company. Its share of the market was simply not big enough to enable it to compete against the two major companies, British Motor Corporation (BMC) and Ford. As a result, it was constantly running short of capital to develop new models. When it did come out with a new car, the financial returns from the sales it was able to achieve were scarcely enough to keep the company in business.

Black had, however, found a way of surviving by manufacturing tractors for Harry Ferguson. He had met Ferguson through their mutual interest in farming and, when Lord Nuffield of Morris Motors had turned down the opportunity to make the Ferguson tractor, John Black had stepped in. The original agreement between Ferguson and Black had been written on the back of a Claridge's menu after a luncheon.

Like all other business arrangements with Ferguson, the relations that developed between the two men were stormy and sometimes hostile. Ferguson was continually carping about the costs of manufacturing at Standard-Triumph's Banner Lane plant, and complaining that his idealized Price Reducing Scheme could not function because the costs to him were so high. In this case, Ferguson did have legitimate grounds for complaint. It was becoming increasingly clear that Standard-Triumph was able to make a profit only because of its tractor business, and that its ten-year contract with Ferguson was effectively subsidizing the car division and enabling it to stay afloat. For Standard-Triumph, the manufacture of Ferguson tractors had become all-important. By the time Massey took over Ferguson, it accounted for 70 percent of profits.

Black had had a go-ahead from his board of directors to try and negotiate a merger with Ferguson himself. The board was therefore surprised, and upset, that Massey made its move first. They then decided, rather recklessly, that they would not make tractors for the new company when the existing agreement expired in 1955. Black promptly ignored this decision, went over their heads and signed a twelve-year contract with Massey.

In the next few weeks, the Standard-Triumph board was to be shocked once again. The firm that made car bodies for the company was sold to BMC. To add to Standard's difficulties, there was a sense of growing alarm about the erratic behaviour of Sir John Black. Something of a martinet, Sir John was unpredictable. He had a habit of attending the firm's Christmas party where he would remark upon the number of managers in the room. The following day he would proceed to draw up a list of those who should be fired. The Christmas, 1953, firing line included a member of the board. He objected to his colleagues, and plans were laid for a palace revolution.

Sir John was at his home recovering from a car accident when the butler announced that a deputation was waiting to see him. Ushering the group into the sitting room, the butler then introduced them: "Sir John," said he grandly, "the members of the board of Standard-Triumph." The spokesman for the group was the assistant managing director, Alick Dick, who had worked as Black's personal assistant and was deeply unhappy about having to demand his patron's resignation. Black asked why they were forcing him out. Dick replied: "If you think, Sir John, you will know why."

Black signed the letter, which attributed the reasons for his resignation to ill-health following his car accident. The new man at Standard-Triumph was Alick Dick. Just thirty-seven years old, Dick was an ambitious and social young man, who was also adept at capturing the headlines and using the latest advertising medium — television — to sell cars.

Massey's interest in these goings on was limited at first. It had taken a shareholding in a French firm, Société Standard-Hotchkiss, set up by France's Hotchkiss and Standard because it saw merit in replacing its ageing Pony tractor with the new Ferguson tractors. The French company needed additional capital, which Massey provided in return for a 25 percent interest. But Massey, still in the hands of James Duncan, had no equity interest in Standard-Triumph in Britain, despite the fact that it was wholly dependent on the firm for its European tractor production. The suggestion had been made that Massey should offer to buy the Banner Lane tractor plant, but for the time being it was turned down.

As it developed, the arrangement between the two companies was plainly unsatisfactory — especially to Massey. Anxious to make his name and develop new cars such as the Triumph Herald and TR 4, Dick announced that basic tractor prices would have to be raised with additional profits taken from this. He also stood in the way of changes and of the switch-over to producing a new, larger tractor. Massey had little ability, and no authority, to control these events. Indeed it could be prevented from coming out with a new tractor if Dick refused to spend the funds needed to re-equip the Coventry plant. A further cause of friction developed when Standard began talks about amalgamating Standard-Hotchkiss (in which Massey had a 25 percent stake) with a French engine manufacturing firm without bothering to tell its partners in Toronto.*

By early 1956, exasperation had grown to the point where the executive committee of the board at Massey were prepared to act. There were a number of things to worry about at home, as the crisis leading to Duncan's resignation gathered momentum. Still, Bud McDougald was asked to begin accumulating Standard stock with a view of improving Massey's position. With its huge new tractor business, Massey could not afford to be held hostage. Unlike Harry Ferguson, the Argus group had the leverage and the capital and were prepared to risk the latter in order to safeguard their interests.

The strategy was to buy into Standard without raising the alarm. If it was known what Massey was doing, then the price of the stock would have risen substantially. To disguise its purchases, McDougald did not deal through Wood Gundy — a firm closely linked with Massey since the 1920s — but with another Toronto investment house, Dominion Securities. By the fall of 1956, he was able to report back that five million Standard shares, about 20 percent of the total, were in his hands. The executive committee then gave approval to Phillips intervening and holding talks with Standard, with the aim of acquiring another five million shares.†

Phillips did not take action immediately. Standard's car business was in poor shape and the company had been forced to lay off 1,000 workers and put the whole of its assembly line staff on a three-day week. It had also been driven into negotiating, first with the Rover Company and then with Sir William Rootes of the Rootes Group, for a merger. Since Massey's interest was only to get control of tractor production, Phillips saw an advantage to be gained by practising patient diplomacy. He would try and smooth the talks along in the hope that an alliance between Rootes and Standard would open the way for Massey to make a cash offer for the tractor plant.

Phillips' tactics were backed up by Massey's senior managers and given an added urgency by a memorandum from Hal Wallace who had looked over the Banner Lane plant. In

* At an early stage in their relations with Standard, Phillips, Taylor, and McDougald — all resolute anglophiles — decided that the company's managers, in particular Sir John Black and Alick Dick, were not the kind of Englishmen they were used to doing business and socializing with. From then on, relations got steadily worse.

† The attempt to influence Standard through the acquisition of 20 percent of the stock was a serious miscalculation. The Argus partners had thought that this shareholding would force management to acknowledge them and their interests, and perhaps gain them control in the way it had with Canadian firms. The British, however, remained unimpressed, and the Standard board made no concessions at all.

164 Wallace's opinion, the plant was "an excellent facility and, unless it is acquired, Massey-Harris-Ferguson will be in almost as vulnerable a position as Harry Ferguson was with Ford." Wallace recommended not merely buying the tractor assembly operation but also the engine manufacturing plant.

That winter, Phillips and Taylor invited Alick Dick to join them in their retreat in the Bahamas. Taylor had begun his development at Lyford Cay and was converting an idyllic stretch of Atlantic shoreline into a rich man's resort, with a yacht harbor, golf course, clubhouse, and luxurious mansions. Phillips had his yacht *Sea Breeze* in Nassau. And, conveniently, Sir William Rootes was spending that winter in the Bahamas. The three men, together with their rich friends, set out to persuade the young, impressionable Dick that it was in everyone's best interest that a deal should be struck.

Dick was impressed. He recalled later that one meeting had taken place aboard an American-owned yacht which "was 380 feet long and had color television, not to mention gold lavatory chains." But he had been joined by another figure who was rather more obdurate and less easy to impress. Lord Tedder, formerly Deputy Supreme Allied Commander under Eisenhower, chief of Britain's air staff, chancellor of Cambridge University, had moved to Standard-Triumph as its new chairman. Dick and Tedder were not won over by Rootes' plans for the new company. One of his proposals was that members of the Rootes family should hold the posts of vice-chairman and president, with three young scions of the Rootes family being managing directors. Tedder would be chairman and Dick one of five managing directors.

Behind Rootes' plan loomed Phillips and Taylor. It had been tacitly agreed between them that when the merger went through, Massey would pick up the tractor plant. The talks dragged on. But ultimately Dick and Tedder refused to agree. Instead they proposed to forget about Rootes and his empire-building, and investigate closer links with Massey. They hoped the Canadians would agree to inject an additional amount of cash into Standard to help revive their car business.

An arrangement along these lines was not what Phillips and Taylor wanted. They had no wish to get entangled in the British car industry and were concerned about the political implications of a group of Canadians seeming to bid for Standard-Triumph. The intent was to get control of the tractor manufacturing operation only. Having failed to push the Rootes bid successfully, however, they felt they had little choice. So it was agreed that Massey would make a takeover bid for Standard through an exchange of shares, and Tedder would recommend to shareholders that they should accept the offer.

On July 17, 1957, it was announced that Massey would offer one share of its common stock plus twelve shillings for each unit of eight shares of Standard. At current stock prices and at the prevailing rate of exchange between the British pound and the Canadian dollar, the offer was worth about $25 million.

Almost as soon as the offer had been sent out, E.P. Taylor — the board member with the most financial acumen — realized that it had serious shortcomings. Basically, Standard shareholders were being asked to make a judgment about Massey's earnings prospects. And Taylor, as chairman of the executive committee, knew all too well that Massey's earnings in 1957 (in the aftermath of Duncan's departure) were going to look dismal.

In a letter to Phillips, Taylor pointed out that Lord Tedder had not helped matters at all. He had told shareholders that Standard was anticipating a substantial improvement in profits, which was exactly what he should have avoided saying if the deal was to be made appealing. While it was true that Massey's losses were nonrecurring ones and that its outlook beyond 1957 was favorable, Standard shareholders would have every reason to be doubtful. No new earnings information had been put out to win them over, though the transaction had been endorsed by two leading merchant banks, J. Henry Schroeder and Helbert Wagg and Company.

The rebellion by Standard shareholders was led by an accountant, and taken up by the financial press. Massey was not offering enough. In view of Standard's underlying asset wealth, the Canadians were trying to get themselves a bargain. The *Stock Exchange Gazette* of London called on the directors of Standard to turn down the offer which had yet to be made formally. And, to add to Massey's embarrassment, an MP rose in the House of Commons to ask whether the government was going to tolerate control over Standard-Triumph passing into the hands of foreigners.

Eric Phillips was not a man to duck a fight. He made it clear that if the offer was not going to be acceptable to Standard shareholders, then Massey would take its business elsewhere and build its own tractor plant, which would, of course, have a devastating effect on their company. This position was interpreted by the British press as being tantamount to blackmail.

Events were, however, running against the deal, and not just because of the hue and cry that had been raised. During the summer, there was a precipitous decline in the Toronto stock market and in Massey's own stock. More was being learned about its earnings. Moreover, the French franc was devalued, undermining the company's position in one of its most crucial markets. By the fall, the value of Massey shares had fallen from $7 to below $6. And, on September 8, the company issued a press release saying that it would not proceed with the offer.

With the takeover called off, a period of fractious relations ensued between the two companies. Neither side could agree on prices for the work that was being done, and the quality control at Standard began to slip, causing more problems. By the spring of 1958, there was almost open warfare between Massey and Standard.

The final break came when Standard sought to acquire an auto body making firm, Mulliners, through an exchange of shares. The matter had been raised by Tedder on a visit to Toronto and had met with strenuous objection from Phillips since the net effect would be to dilute Massey's holdings in Standard considerably. Far better, said Phillips, to allow his group to purchase more shares in Standard and then make a cash offer for Mulliners. This way, the acquisition would be achieved at a better price.

Phillips presumed he had an agreement, but Tedder as a military strategist had other ideas. Perhaps he sensed that it was time for a final skirmish with Massey. When Phillips learnt about Standard's offer to Mulliners, he was furious; so furious that he devised a counter-offer under which Massey would buy from Mulliners shareholders all the Standard shares that they would receive at a price that was above the current market price. Tedder publicly repudiated this action. "The Board of Standard regret this unilateral action which they feel is inconsistent with the mutual confidence which is so essential to the interest of both companies," ran the text of a prepared statement. It then went on to state that Massey was plainly making a grab for control of Standard, and that this would be resisted.

It seemed that Tedder and Standard were the winners in this duel. Aside from making its counter-offer, Massey did not release any more information or attempt to explain what Phillips considered a betrayal of his understanding with Tedder. Much of the British press was hostile to the Massey action and regarded it as unwarranted, one newspaper even referring to it as the behavior of "a spoilt child."

Phillips, however, was not a man to lash out blindly or to take action out of mere pique. He recognized that relations with Standard had already deteriorated to such an extent that Massey would have to step in and buy the tractor plant. He therefore made the break inevitable. At the same time, he moved behind the scenes to buy a larger interest in Standard-Hotchkiss, ensuring an equal partnership in the French tractor manufacturing operation, and he approved negotiations for the acquisition of another British firm, F. Perkins Limited of Peterborough, so that Massey would not be dependent on Standard for engine manufacturing facilities. These two moves gave Massey a far stronger hand to deal with Standard. The acquisition of Perkins

Massey could move to set up its own tractor plant and cut them out altogether.

Phillips' actions forced Standard's management to be realistic. Dick came to the conclusion that it was better to sell out while the contract with Massey still had seven years to run. Moreover, the cash such a sale would bring in was desperately needed. Eventually he sat down with Thornbrough and they began to haggle over terms. Massey's negotiating team, which included John Staiger and Hal Wallace, expected that the negotiations would take about ten days; instead they dragged on for four months.

By the time negotiations got underway, all semblance of co-operation and trust between the two sides had broken down. The Standard team wanted to exact the maximum amount of money from Massey; the Massey team, suspicious about how Standard's tractor and automotive divisions would be prised apart, insisted that everything must be listed and that there must be penalties if Standard did not live up to the deal.* Further questions arose over language. Having negotiated all week at the Standard office in Berkeley Square, Kenneth Aspland, the company's finance director, would return to Banner Lane over the weekend. Arriving back in London on Monday he would announce that the Standard board had rejected the language of the proposed contract, in particular the "American" phraseology of Massey's negotiators: Everything must be couched in proper English.

By the end of July, 1959, after protracted discussions and bitter exchanges, an agreement was finally arrived at. For $32 million, Massey would buy all the tractor assets of Standard in both Britain and France, including the plant in Banner Lane, Coventry, which by itself had the capacity to manufacture 100,000 tractors a year. The Massey interest in Standard amounting to 7,758,000 shares would be bought out for $8.3 million. In France the company would acquire both the Saint Denis and Beauvais plants of Société Standard-Hotchkiss.

As it emerged, the deal seemed an expensive one. Having acquired Perkins, Massey did not take any of Standard's engine manufacturing facilities, which it did not need. What Massey wanted, and got, was ownership of a giant tractor manufacturing plant in Europe and, with it, far greater control over pricing and profits. It was an important step towards consolidating the company's position.

As to the months of hard bargaining with Standard and the suspicions of bad faith, these were at least partly confirmed

* The itemized list included fifty-four pigs, with a brief description of each pig including its color and sex, and was entered on the deed of sale. The pigs were being offered to Massey as part of the garbage disposal unit that serviced the factory canteen.

when Massey executives moved into the new offices in Coventry. The typewriters on the desks, which had been included in the contract of sale, were all there. And so were the staplers. But the Massey negotiators had failed to specify that the typewriter ribbons and the staples were included in the purchase price. Before leaving the office on the previous Friday, the Standard employees had been ordered to take the ribbons and staples with them.

The purchase of Perkins.

The struggle for control of Standard's tractor business had been troublesome and exhausting.·But the goal had eventually been achieved, and Massey was, in the words of Eric Phillips, a master in its own house. Now Massey was able to manufacture its single biggest selling item under its own terms and with its own work force. Tractor production was no longer dependent on a firm that had its own shareholders' interests, not those of Massey, to consider first.

Still, the long-fought duel had not been without cost. Massey had had to pay dearly for its tractor purchase in comparison with other acquisitions it had made. And the unfavorable publicity that had swirled around its attempts, first, to take over Standard and, second, to stop Standard taking over Mulliners, had dealt some blows to its credibility. None of these setbacks, however, was to prove crucial. Further, the handling of the acquisition of another old-established firm, F. Perkins Limited, did much to redeem Massey's reputation.

The purchase of Perkins was everything the Standard deal was not. It was conducted amicably and quickly; the actual talks lasted only six days after it had been revealed that Massey was interested. The whole move had been planned in advance, and relations between the two companies had been carefully built up. There was no rancor and no major differences to resolve.

For what was to be a bargain price, Massey obtained not only a diesel engine plant, but a firm with a bright future independent of the farm business; Perkins had its own customers and a thriving international business, with subsidiaries and licensing arrangements in a number of countries including Yugoslavia, India, France, Brazil, Spain, Argentina, and Italy.

One of Perkins' customers was Ford, and there was concern in Toronto that the Ford management (Ford was a competitor

in the tractor market*) would never allow the takeover of
Perkins. In fact Ford chose not to act. As a measure of the coup
that Massey was able to pull off with this acquisition, Henry
Ford II was later to remark that standing by and allowing the
sale of Perkins without having put in any kind of counter-bid
was one of the worst decisions his firm ever made.

The purchase of Standard's tractor plants represented
the finalization of a piece of unfinished business from the past;
the legacy of Harry Ferguson and his cavalier attitudes to
business integration. Having become a major distributor of
tractors, indeed the biggest in Europe, Massey simply had to be
a manufacturer as well. The Perkins deal was different. Not
a defensive move in any way, it laid the base for the company's
moving ahead in the 1960s and 1970s; for Massey's expansion
into new areas and new territories.

Frank Perkins and Charles Chapman were two gifted
engineers who had developed a light, high-speed diesel engine
in the 1930s. They had first worked for other companies, then
founded their own firm in Perkins' home town of Peterborough.
The engine they developed was intended for use in commercial
vehicles, and so they undertook conversion work switching
over trucks and vans to the more economical diesel engine.
Business was not brisk. For many years F. Perkins Limited of
Queen Street, Peterborough, survived only because it worked
out of premises owned by Frank's father and was bankrolled
by a family friend.

The breakthrough came in 1937 when the firm's two
founders came out with an efficient six-cylinder engine, the
Perkins P6. This set the stage for a period of rapid advancement
as the new engine was sold for use in boats, trucks and vans,
and industrial machinery. Both materials and money for
expansion were hard to come by during the war, and immediately
afterwards. But the company borrowed from the government,
started up its own factory outside the town of Peterborough
and, by 1951, had done well enough to launch itself through
a public stock issue. In 1946, the young company made just
£10,000 profit on sales of £1.25 million. By 1954, it was earning
£432,000 and sales had soared to £16.8 million.

What Perkins had done successfully was to perfect one or
two engine models, particularly P6, and sell them to vehicle
and tractor manufacturers. It had established subsidiaries abroad

---

* In this period, competition from Ford was strong in the tractor market. In North
America, a new series of tractors put out by Ford included one which had greater
horsepower than the MF-35; because of this, Thornbrough and Klemm hurriedly
designed and put into production the MF-65. In Britain, Ford and Massey were close
competitors, sharing between them about 80 percent of the tractor market. At first,
management in Toronto measured the success of the company against Ford; later,
after Massey surpassed Ford in farm equipment, the target became John Deere.

170 to sell and service its engines. And it had made arrangements for Perkins engines to be made under licensing agreements (in India, Yugoslavia, and France); through a minority position in a local firm (in Spain); and by pledging to set up its own plant (in Brazil).

All this had worked well for a time. Perhaps too well. Like many technically minded people, Frank Perkins had been content to see his design ideas come to fruition and find a market; he had not concerned himself with the constraints that might affect his firm once it became a supplier to others. In fact, there were limits to the growth that could be expected.

Having adopted the diesel engine, Perkins customers were prepared to stick with the firm at first. But as their requirements became larger, and Perkins more successful, they began to think of manufacturing for themselves and producing their own diesel engines. Perkins had been carried away by the success of its P6. Not enough had been spent on research and development, and other companies were beginning to come along with lighter-weight, more economical, engines. Sensing that the firm was in danger of being left behind, a new design, the R6, was hastily developed and put into production. It had not been adequately tested, and complaints started streaming in. Perkins was in trouble. To add to its difficulties, it suddenly lost important markets in India and the Far East because of the Suez crisis and the blocking of the canal.

The switch from profits to losses came quickly. In 1957, the value of sales fell by nearly one-third and earnings of £348,000 the year before turned to losses of £319,000. Frank Perkins' confidence was shaken. He turned to his second-in-command, Monty Prichard, and asked him to take over the running of the company and prepare a report for the shareholders on what should be done.

The Prichard report was, like its author, a decisive and plain-spoken document. It pointed out that the company's profit performance had been slipping even in the good years, and that it was running short of customers; three vehicle manufacturers had been its principal clients a few years earlier but now each of them was making its own diesel engines and ordering none from Perkins. The company needed new designs, but it could not afford the development costs and, in its shaky condition, could not hope to raise financing. Prichard was adamant. To stay in business, Perkins must have competitive products, and this meant it would need an infusion of capital beginning in 1959. The only way vital capital could be obtained was through "amalgamation with a complementary organization with cash available."

The first approaches were made to the car companies but

they were not interested. Prichard then raised the matter with 171
Eric Young, Massey's managing director of Eastern Hemisphere
operations. Contact between Perkins and Massey had been
made at a social level — Frank Perkins and Guy Bevan were
friends of long-standing. The two companies had done business
together. Five years earlier, Perkins had hoped to get a major
engine order for Massey's planned European tractor, but the
contract had gone to Standard.

Prichard was due to make a trip to Brazil where Perkins
was building a plant; on the way back, he came to Toronto and
met with Thornbrough.* Massey, he was told, was interested in
doing a deal quickly and emerging with full ownership of
Perkins. Within a week Prichard was back in Peterborough and
the deal was agreed; Massey would make an offer for all the
issued Perkins shares worth about £5 million. This time there
was no friction and no fuss. For what in retrospect proved to be
a ludicrously cheap price, Massey had obviated the need to
concern itself with engine supplies and bought a source of its
own, and it had acquired an extraordinary asset. Henceforth,
Massey tractors could be manufactured in Britain or India,
France or Brazil, and powered with Perkins engines made in the
same country. It would be a far more cost-efficient and profitable
way to do business.

The "three-legged stool," an international gameplan.

In the space of little more than six years, Massey had bought
the business of Harry Ferguson, dealt for Standard's tractor
operations and purchased Perkins, three acquisitions that
changed the makeup of the company internally, and led to its
emergence as a major force in British industry. Viewed from
Toronto, the push was not justified by the British market
alone, rather it was part of an international gameplan in which
Britain was to be the base. The company was engaged in raising
its profile around the world, expanding internationally to
compete with its rivals who held their secure markets in the U.S.
corn belt.

In 1958, Massey assets employed in Europe had been half

* The first overtures were made when Prichard came to Toronto to attend the automotive show in October and talked with Thornbrough and Staiger about the possibility of Massey acquiring Perkins. This was followed by an undercover visit to the
Perkins plant by Staiger and Wallace. Their recommendation was positive. The idea
was then raised with Phillips and McDougald who saw an immediate advantage to be
gained; the acquisition of Perkins would jolt Standard into negotiating seriously in
London.

the figure for North America — $96 million versus $183 million. Three years later, having absorbed Standard and Perkins, the relative positions were almost equal; assets employed in Europe had jumped to $228 million against $234 million in the U.S. and Canada. Moreover, this advance in Europe was accompanied by a great build-up elsewhere. Between 1958 and 1962, the tally of assets employed in Australia, Africa, and South America more than doubled to $60 million.

Massey's growth was being built on a solid base. Originally, the expansion into Europe had been seen as a counter-offensive, a way of getting around the weakness of Massey — and the strength of its competitors — in the U.S. But world markets for farm machinery were in transition and, by 1960, it was evident that farmers in Europe had a greater need for new farm machinery and for mechanization. In the U.S., the market was now largely supported by replacement orders, and by an increasing trend towards gadgetry and gimmicks, which were not always profitable for the farm equipment companies. Unwittingly, perhaps, Massey had got itself out ahead.

In the U.S. market, Massey was battling for third place with Allis-Chalmers and Ford, well behind John Deere and International Harvester. Outside North America, it was the leader, outstripping even International Harvester, its nearest rival. But more significant than the company's market position was the greater opportunity that overseas markets held out for future growth, and for profitability. Farming in the U.S. had been in decline during the 1950s, with fewer farms and larger amounts of capital being invested. More and more, American farms took on the characteristics of big business. For the farm equipment makers, this meant the erosion of the mass market for their products. Farm equipment manufacturers had to come to terms with fewer farmers wanting fewer machines to do a bigger job.*

One reaction to this was to offer more comfort and convenience to the American farmer, upgrading farm machines as if they were automobiles. John Deere scored a hit by introducing a tractor seat designed by Dr. Janet Travell, the doctor who had recommended a rocking chair for President Kennedy. Still, the emphasis was on size and performance.

Since competition was keen and profits small, the sudden downturn in sales experienced in 1960 put several companies into financial difficulties. To try and raise their share of business, they were forced to spend extra money on promotion, often

* Figures from the U.S. Census of Agriculture showed that between 1954 and 1959, 850,000 farms went out of business, a decline of 18 percent. At the same time, the average size of those farms that survived expanded to 302 acres from 242 acres, and the amount of capital investment in a farming operation grew substantially; by 1960, it was commonplace for total farm investments in the corn belt to be $100,000 or more.

with poor results. Deere took over the state fairgrounds in Texas
and flew 6,000 dealers to Dallas for a sales extravaganza (They
were serenaded by a company promotional song which went:
"There'll be new demands for power/Like nothing you've ever
seen/A whole new kind of production/That's gonna move,
really move, I mean"). One thing that did not move was Deere's
earnings which dropped to a new postwar low. Marc Rojtman,
the energetic president of J.I. Case, decided to whip up sales
by inviting thousands of dealers to Nassau and Miami Beach to
be wined and dined. One year later, Case was being bailed out
by a consortium of banks, alarmed by its $178 million debt, and
Rojtman had lost his job.

The crisis served to underline the fact that the industry had
become a marginal business. Major investments had to be made
in manufacturing and distribution, especially for the large
firms which offered a full line of tractors and implements. The
usual ratio was $1 of current assets to produce $1 in sales.

Liberal credit terms were extended to dealers and farmers.
The manufacturer might have to carry a dealer's inventory on
the sales floor for a period of one year or eighteen months, and
possibly also carry the trade-in equipment that the dealer
accepted to make a sale. So, having made the delivery to the
dealer, the manufacturer would be faced with receivables on his
books until the final sale to the farmer, possibly much later.
This way of doing business was not liked, but in a competitive
industry it could not be changed. Moreover, retail credit terms
were exceedingly generous. A farmer might pay for a purchase
over four of five crop years. Sometimes he would be given a
cash payment as an incentive to place an order.

All of this meant that a temporary dip in sales had a sustained
and serious impact on the farm equipment companies. The
period that elapsed between manufacturers shipping their equipment and an actual sale was extremely costly. And the costs
were borne by the industry.

These terms of business had a debilitating effect on profits
even for the industry's bigger companies. And, since the North
American market was turning down anyway, they had reinforced a trend towards diversification. When the chiefs of the
industry looked around they saw two ways of improving
things. One was to move into new products in a related
area, the favorites being industrial and construction machinery
and trucks. The other was to expand abroad where markets
were promising and sales terms less onerous.

As the 1960s opened, International Harvester was drawing
more of its sales from industrial equipment and trucks than
it was from farm equipment. And Deere, traditionally the most
stay-at-home company in the industry, was taking its first profits

from a German subsidiary that it had acquired four years earlier, and had just announced that it would build a tractor plant in France.

Massey had by then made its moves. It had positioned itself in the still-developing farm markets of Europe as well as in the overdeveloped markets of North America. And it had long-range sales ambitions for Africa, Asia, and South America.

It had also diversified into new products with the acquisition of Perkins, a supplier of diesel engines to many other customers apart from Massey. Moreover, it was rapidly building up Perkins, providing it with the capital funds and financial support it had lacked in the past. At the time of the takeover, Perkins had been producing 77,000 engines a year. By 1963, output was up to 250,000. Perkins under Massey, and under the leadership of Monty Prichard, had done a remarkable job of holding on to old customers and finding new ones.

Another element in Massey's plans for diversification was to move into industrial machinery. This had started out through the purchase of Mid-Western Industries, a small company in Wichita, Kansas. Mid-Western's equipment consisted mostly of tractors, with loaders and back-hoes. In the early 1960s the manufacturing operations for this were switched to Detroit. An attempt was then made to expand into industrial and construction machinery; this was to be the third leg in what Thornbrough referred to as Massey's "three-legged stool" (farm equipment diesel engines, and industrial and construction machinery or ICM), and hopes for success were high. Profit margins would be better than for farm machinery, according to forecasts made at the time. By widening its base still further, Massey would be following the corporate trend of the 1960s, creating new divisions and a new earnings base.*

As it turned out, the move into ICM was to prove a major disappointment. There was a continual debate about how far the company should venture into this new area, dominated by firms such as Caterpillar and Japan's Komatsu. Efforts were made to upgrade Massey's line of products. When an Italian company, Landini S.P.A., was acquired in 1963, a new line of crawler tractors was added. And a top marketing man, John Mitchell, was recruited from Clark Equipment to run the ICM division. Still, management was torn between increasing its commitment or holding back. And, when in the early 1970s, it finally came down on the side of a major expansion, the

---

* By 1966, the year in which the company set up its separate Industrial and Construction Machinery Group, ICM sales were $75 million against total sales of $932 million. Farm machinery sales accounted for $668 million; engine sales for $99 million; and the sale of parts for $91 million.

consequences were to prove disastrous.   175

At the time, ICM was not seen as a make-or-break endeavor. It was simply another indication that Massey was moving with the times; it showed the company was well-managed, and responsive to the opportunities that seemed to exist in the 1960s. In developing its three-legged-stool policy, Massey was doing what other multinational companies were doing, striving for greater growth and greater stability of earnings during a period when business was expanding almost everywhere.*

Certainly, the organization and structure of the firm put emphasis on growth and on aggressively seeking out new places in which to do business.

Massey was structured into separate manufacturing areas, or operations units. Each of these were profit centres responsible to the president in Toronto. Grafted on to them was an export organization, which acted as a trade agency and sold to 150 countries where Massey did not have its own plants.

With the creation of a Special Operations Division, Massey had moved a step ahead in its planning for the future. Located in offices close to Berkeley Square in London were a group of executives charged with masterminding the firm's progress in the international arena. The division's thirty members were led by two Englishmen, John Beith and Len Boon, and their job was to investigate places for Massey to expand and set up manufacturing operations. Early locations that were being considered were India, Yugoslavia, Brazil, Italy, Turkey, Spain, and Argentina. In each case, the Massey team would make its assessment, and if the recommendations were favorable and were acted upon, the investment would take place and a new operations unit would come into being. An example of this was in Italy where Perkins had supplied diesel engines to the Landini Company. A decision was made to buy Landini, keeping its name and product lines, and it was then brought into Massey as a fullfledged operations unit.

By inclination, and by design, the thrust was towards expansion and towards generating new profit centres and a maximum amount of global activity. Much of this emanated from the top. Thornbrough was ambitious to see Massey outpace and outperform John Deere. Moreover, with Phillips still at the helm as chairman, there was little fear of the company overreaching itself or getting into financial difficulties. The expansion that was taking place made sense, and it was generally

* In recalling this period, Thornbrough says that diversifying and creating new divisions was almost a religion: "In the sixties, if you had exploited the opportunities in one area, you simply had to give your managers and shareholders a chance to get into something new."

176 well-planned and co-ordinated.*

The greatest success of all, and the place where the growth strategy paid off most handsomely, was Britain. By the early 1960s the country had become a focal point for the industry, and for global sales; as early as 1960, Britain had produced more tractors than the U.S., (190,000 versus 146,000). In the British market, nearly half of all tractors sold carried the Massey-Ferguson emblem, while in combine harvesters Massey had four-fifths of the market all to itself. The use of Britain as a base was paying dividends. Massey's great sales strength was in the countries of the sterling area, and it was in these countries that it was the undisputed leader.

For the company itself, these new directions had inevitably brought changes. In the Duncan days, both management and ownership of Massey had been resolutely Canadian. In the post-Duncan period, the Americans had taken over to introduce new organizational methods and skills, and to turn the company around. After 1959, and the string of British takeovers, a fresh element was added; British executives became more prominent, among them Tim Powell, in charge of Massey's British operations, Monty Prichard at Perkins, John Beith and Len Boon at Special Operations, and Peter Wright in the Export Organization.

The influx gave a distinct character to Massey. Here was a company that remained in the grip of a select group of Canadian millionaires, with its board made up of themselves and their wealthy friends. It was managed by Americans who held on to most of the senior positions and initiated policy decisions. At the same time, much of the dynamism and forward momentum was coming from Europe, where sales and earnings were being generated, in London, Coventry, and Peterborough. For a time, in the 1960s, this blending of talents and personalities worked extremely well. Massey-Ferguson was the model of what a progressive, multinational company ought to be, and was widely admired as such.

* By 1964, the year of Colonel Phillips' death, Massey-Ferguson had subsidiaries in Argentina, Australia, Brazil, Canada, Eire, France, Germany, Italy, Mexico, South Africa, Southern Rhodesia (Zimbabwe), the United Kingdom, and the United States, and plants in all these countries except Agrentina, Eire, and Mexico. Its international group operated out of Switzerland, Panama, and the Netherlands Antilles. And it had an associate company in India. In addition to this, the Perkins Group based in the United Kingdom had subsidiaries in Australia, Brazil, Canada, France, Germany, Italy, South Africa, Spain, and the United States.

11
Towards the Brink

Massey seemed to have succeeded in attaining almost all its goals. But with that success had also come a sense of detachment on the part of management, and of isolation, a willingness to sanction new expenditures and take on new commitments with almost cavalier abandon.

By the early 1960s, the winter homes of two of Argus's four partners were situated on the same stretch of expensive shoreline in Florida. South Ocean Boulevard follows the coast, skirting the municipal beach and then heading out of Palm Beach, with the sea on one side and a panorama of gilded mansions on the other. About one mile south of the town are two Mediterranean-style villas, painted white with red roofs and surrounded by high walls; two narrow roads — one called El Bravo Way and the other El Brillo Way — intersect the homes.

In 1960, Bud McDougald had bought the home on El Bravo from a member of the Du Pont family and had established the mansion as his winter headquarters. Between November and April, he was to be found in Palm Beach. During these months, he took an active part in the social life of the community, taking afternoon tea at the Breakers Hotel and mixing with the rich and well-connected at Palm Beach's two exclusive clubs, the Everglades and the Bath and Tennis. Here he entertained old friends from England who were invariably titled and aristocratic, and indulged himself in his favorite pastimes of deal-making and doing business; for McDougald, the relaxation and pleasures of Palm Beach were not an excuse for a personal diversion. He was there to see people, and be seen by them, and to run his global business interests from the palatial splendor of his home.

An important part of McDougald's dealing involved Massey-Ferguson. Whether he was at Palm Beach or his suite at Claridge's Hotel in London, Massey and its affairs were of particularly close interest to McDougald. It was his association with Massey more than with any other company that established McDougald's credentials with the landed gentry of England and the top men of the global business community. While Eric Phillips was alive, McDougald was just one member of a six-man Executive Committee of the Board that met monthly. Even so, during the winter months, the committee would often meet in his Palm Beach villa.

It was after one such meeting that Doris Phillips announced that, unknown to the colonel, she had purchased the house next door to the McDougalds on El Brillo Way. Doris and McDougald's wife, Jim, were sisters, so the move was not altogether unexpected. But it came as a surprise to the colonel. For many years he had retreated to a winter home in the Bahamas and, in the postwar years, he had introduced his new business partner, E.P. Taylor, to the islands and to Lyford Cay. Taylor had proceeded to start up his own business and to found his land development company to build homes there. This had not in any way affected the colonel until the two of them had started having differences and began to drift apart. In the Bahamas, Phillips now chafed at his close proximity to Taylor. Sensing this, his wife had initiated the move from Nassau to El Brillo Way.

With the colonel and McDougald in Palm Beach and Albert Thornbrough, Massey's president, having a home thirty miles away in Boca Raton, the Florida coast became a winter refuge for the men who presided over the affairs of Massey-Ferguson. Florida also became symbolic of something more. The relationships at Argus that had existed at the time of Duncan's dismissal, and the kinship then between Phillips and Taylor, was now a thing of the past. To this was added the cleavage that had always existed between McDougald and Taylor, and the fact that the fourth member of the group Wallace McCutcheon was losing interest. McCutcheon was soon to take himself out of Argus altogether. In the 1950s and early 1960s, the Argus connection had underpinned Massey's efforts to expand, and helped the company considerably. But gradually, it was turning into something less constructive and less useful.

While Colonel Phillips was alive there was not much to worry about. In the years of Massey's expansion, the support of Argus and the presence of Phillips counted for a great deal. Among the other Argus-controlled companies, none commanded the same degree of attention or personal involvement as Massey. Phillips had an office at the King Street headquarters. When in 1961 the company moved out of the crumbling offices that Hart Massey had built to new premises on University Avenue, the colonel, now in his late sixties, moved too.

By this time Phillips was in failing health and he was   179
spending less time in Toronto. But he retained the post of chief
executive officer as well as chairman, and continued to involve
himself in the affairs of the company when he could, making
trips to Europe, chartering a train to take his friends to the
opening of the new combine plant in Brantford, Ontario, and
appearing at dealers' shows and at presentations given to
financial analysts.*

Over the years Phillips had put the reputation of Argus
behind Massey and this had served as a guarantee in Massey's
business dealings. In addition, the relationship that had developed
between Phillips and Thornbrough was a cordial and constructive
one. In running the company and shaping the strategy that
was taking Massey into new areas of manufacturing and new
locations, Thornbrough was given full backing by Phillips.
As an internationalist, the colonel was as convinced as Thornbrough that expansion must be ambitious and far-ranging
if the company was to match its competitors. And as a businessman he could appreciate the results that were being obtained.

In the early 1960s, the direction that Thornbrough was
taking proved extraordinarily successful. The expansion that had
taken place had given the company new industrial capacity and,
by a happy coincidence, this had occurred at a time of worldwide growth. In North America, the farm market had bounced
back from the downturn of 1960. Farmers were presented with
a fast-growing global market in which food needs were running
ahead of production. Crop prices were high. And farmers were
flocking into dealers' showrooms to buy the latest equipment.

For Massey, which was also able to exploit its new strength
in tractors and diesel engines, 1960 to 1965 were vintage years.
Over the five-year period, sales advanced at an annual rate
of 11.6 percent while the average annual growth in earnings
was an astonishing 30 percent. In 1966, the end of his first decade
as president, Thornbrough was in charge of a company with
sales of $932 million and profits of $45 million (compared with
sales of $390 million and a *loss* of $4.7 million in the first year
after Duncan's resignation). All this seemed to indicate that a
policy of nonstop, headlong growth had paid off handsomely.

Certainly, the controllers of the purse strings, Argus
Corporation, had no reason to complain. In 1958, the four leaders
of Argus, Taylor, Phillips, McDougald, and McCutcheon, had
bought themselves a magnificent new head office at 10 Toronto
Street, from which to run their corporate domain. The building

* Despite his failing health, the colonel insisted on getting out and showing the flag.
At a distributors' show in Greece, thousands of dollars in unsigned travellers cheques
were found lying on the ground inside a tent. Later in the afternoon, the colonel
discovered he had lost his money.

had been vacated by the Bank of Canada and was threatened with demolition until the four financiers had stepped in to save it. Its imposing Grecian columns were to become a symbol of the Canadian business establishment.* Inside, it was refurnished with opulent good taste, with upstairs offices that were heavily broadloomed and illuminated by crystal chandeliers, and a panelled boardroom downstairs that resembled — to one news-paper reporter's eye — a "baronial dining room."

Argus by now had established itself as the nation's most influential investment company. Its holdings were concentrated in seven major corporations; Massey-Ferguson, B.C. Forest Products, Canadian Breweries, Dominion Tar and Chemical, Hollinger Consolidated Gold Mines, Standard Radio, and Dominion Stores. The market value of the Argus portfolio, which had been $15 million in 1945, was now $125 million. But the power and prestige that surrounded Argus was not merely a reflection of its business success or the control it exercised, it also centred around the personalities, and individual magnetism, of its directors.

In public life, the most active of all was Taylor. His interests and accomplishments were so varied that, at any one time, it was difficult to pin down exactly what his main preoccupation was. In the early 1960s, he was attempting to forge a new brewery empire in Britain, build a luxury development at Lyford Cay, establish Toronto's O'Keefe Civic Centre, and raise funds for the city's General Hospital and its art gallery. McCutcheon, the lawyer, was to leave Argus to take a cabinet post in the Diefenbaker government and then a seat in the Senate.† McDougald, who was by now fully co-opted into Argus, was the negotiator and deal-maker. Although the characters of these men were at variance, they were able to coalesce around Eric Phillips, the oldest member of the group.

In the 1960s, the Argus principals owned around 12 percent of the Massey shares, making them the largest single voting block. The influence Argus exercised over Massey was un-questioned. It was based not only on Argus's control of the board and the executive committee, but also on its handling of Massey's financial affairs and its relations with major lending institutions particularly the Canadian Bank of Commerce.

* Originally designed as the city's seventh post office by architects Colonel Frederick Cumberland and William Storm, 10 Toronto Street had been built in 1853 and modelled after the Temple of Minerva, the Roman goddess of handicrafts.

† McCutcheon did not inform his Argus partners about his political plans. In the summer of 1962 after he had accepted Diefenbaker's offer to become a minister with-out portfolio (later he was made minister of trade and commerce), he telephoned Phillips at his Massey office to tell him. Phillips and the other Argus partners were so irate that when McCutcheon later tried to resume his business activities, they would have nothing to do with him.

For the time being, this was beneficial. Phillips, the strongest member of the Argus group, had a personal interest in Massey; he would see to it that management was backed up with financial support, and he had the connections to help ease the way. If there was a political problem, the colonel would put through a call to his friends in Ottawa; if there was a financial need to be met, then Bay Street's bankers and investment dealers would know about it. As for Massey's forays into Britain and other locations, it so happened that Phillips and his Argus friends could be extremely helpful outside Canada as well, for their influence extended far and wide.

When the Standard deal was being put through, it was Phillips and Taylor who did the bargaining in Nassau. And when Massey needed some blue-chip British directors to add to its board and establish its credentials, Bud McDougald went to work. Drawing on his connections as a member of London's most prominent social clubs and a denizen of Britain's race-tracks, McDougald invited two close friends, the Marquess of Abergavenny, chairman of Ascot, and Lord Crathorne, a former British minister of agriculture, to join the board, which they did in the early 1960s.*

Argus Corporation, and its celebrated cabal of leaders, were at the height of their influence and fame. They dominated the Canadian business scene and everything they had done, or attempted to do, seemed to have succeeded brilliantly. Among the brightest creations of Argus (the name in Greek mythology denotes an ever-watchful beast with a hundred eyes) was the new-look Massey-Ferguson, with its impressive global connections. Phillips had nurtured it along and, in a real sense, its successes were an achievement of Argus, for it was the trio of Phillips, Taylor, and McDougald who had installed the new management, put the deals together, and helped fashion Canada's first truly world-competitive company.

In 1964, the achievement seemed to be a very real one. By the following year, however, it was to look far less substantial. The death of Eric Phillips was to change things substantially. On the day after Christmas, at the age of seventy-one, Phillips

---

* An example of McDougald's ability to pull strings in Britain was the award of a knighthood to Monty Prichard of Perkins. It had been decided that Tim Powell, not Prichard, would become managing director of Massey-Ferguson Holdings, the top job in London, but there was some concern about how the ambitious Monty Prichard would react to this appointment. To mollify him, McDougald suggested he would contact his high-level friends and see if Monty — who was active in promoting British exports — could be included in the Queen's honors list. The knighthood duly came Prichard's way, but he had found out about McDougald's behind-the-scenes dealings, and was angry about it. McDougald, in turn, was miffed by Prichard's lack of gratitude. A luncheon was arranged between the two of them in the hope that things could be smoothed over, but Prichard refused to thank McDougald or make any reference to the event.

182 suffered a fatal heart attack at his winter home on El Brillo Way.

The loss of Phillips was a big blow for Massey though, at the time, many of the company's executives did not recognize it as such. Despite their differences, it was also a blow for Taylor who had lost his senior partner. When in the following year, trouble erupted at Canadian Breweries — the company that had been his personal contribution to the Argus partnership — Taylor began for the first time to question his own ambition. With Phillips gone, did he wish to shoulder the burden of Argus and its dependencies? Would it not be better to trim his own interests, and concentrate on the things he really wanted to do (which happened to be in Britain and the Bahamas rather than Canada)?

Taylor and McDougald had worked together, and benefited from each other's deal-making skills. They had added greatly to their own personal wealth by buying stock through their investment firm, Taylor, McDougald and Company, and selling it to Argus — a legitimate ploy in the early 1950s. But their relationship was not built on mutual confidence or trust.

Now, following the death of Phillips and his own sixty-fifth birthday, Taylor announced his partial retirement. He stood down as chairman of Canadian Breweries and Domtar, though he continued to be involved in Argus for the time being. In 1967, Taylor moved to the Bahamas, and Argus — which had been in limbo — got itself a new president, Bud McDougald. Taylor took over as chairman, a post that had been vacant since Phillips' death. But he stayed only until his seventieth birthday. When he gave up, McDougald took over as both chairman and president.

These changes in the management of Argus were mirrored at Massey with even more unsettling results. Under Phillips, the company had been used to firm and strong direction. From Taylor and McDougald, it got far less. Taylor did not wish to play as full a role in Massey as Phillips had done, but he was content to continue as chairman of the executive committee even if he seldom turned up for committee meetings; indeed, he clung to this position after he had resigned his other chairmanships at the age of seventy. The one thing Taylor did not want, it seemed, was for McDougald to be chairman. For his part McDougald was not interested in managing companies or in the chores of administration, but he was utterly opposed to Taylor being the chairman. As a result, Massey had no chairman.*

---

* The deadlock became apparent at once. The board met after the colonel died and a few weeks before the annual meeting. When it became apparent that neither Taylor nor McDougald wanted the other one to chair the annual meeting, a company bylaw had to be invoked under which, in the absence of a chairman, Thornbrough would preside.

Within the company, the absence of a chairman did not appear to matter much. Thornbrough inherited one of Phillips' posts, that of chief executive officer. Taylor headed the executive committee even though he was not much interested and rarely inquired about what was happening at Massey, and later on Bud McDougald took over. So there was an element of continuity. Nonetheless, the feuding over the chairmanship was more than merely symbolic. It signified a far less responsive approach to Massey, and a general air of disinterest on the part of Argus. Taylor and McDougald were a part of the decision-making process at Massey but they were seldom forceful in pressing their views except in the financial area. Neither of them worked at any time on company business and there was a continual problem about who should be consulted, and about what. Though McDougald was more visible and more involved, he stood back for as long as Taylor headed the executive committee. Moreover, he was busy with other matters. McDougald prided himself on his ability to keep moving, to do deals on the telephone, to run things from Palm Beach and the New York Yacht Club, and his suite at Claridge's; he was not a man to be tied down with problems of detail.

This new and looser relationship between Massey and Argus developed slowly, and seemed to have little bearing on the company's performance for a time. As long as Massey was doing well, it seemed to be of no consequence. But later on, as the company once more foundered, the source of many of its troubles seemed to be traceable to the policy of benign neglect that Argus practised after 1965.

One hundred and sixty-five countries.

The farm machinery industry has always been a demanding one, and sometimes ruthlessly so. The fortunes of Massey had demonstrated this down the years. Frequently the firm had appeared to be on the verge of a period of extraordinary prosperity only to encounter unforeseen setbacks, and find itself once more struggling to survive. It had happened with regularity. In the 1920s, there had been the trend to protectionism around the world; in the 1930s, the Depression; in the 1950s, the downturn in farm markets that followed the ending of the Korean War. Only in the 1940s — when Massey was operating as a producer of arms and munitions as much as farm equipment — had the company continued flat out, selling all the products it could make.

The setbacks that had taken place were not always caused

by events beyond the company's control. Often poor planning, and a failure to respond to changes in the market, had been as much to blame. The troubles of the 1920s, 1930s, and 1950s were made worse by an indifferent and sometimes chaotic approach to managment and by operational inefficiencies. Massey was made to pay for its mistakes. Still, the basic causes of the company's difficulties were rooted elsewhere, and they affected other farm companies as seriously and as often. The cyclical nature of the agricultural business presented its own particular set of problems.

In the production of food, there have been periods of plenty and periods of shortages and high prices. The cycle has not changed greatly since biblical times when the people of Egypt enjoyed seven prosperous years followed by seven lean years. Today, the situation is possibly less desperate since grain reserves are now shared on a global basis. But the underpinning of farm equipment markets continues to be the farmers and their ability to sell their crops when they are harvested. If surplus conditions force farmers to hold back, then they will limit their purchasing. And when farmers are not in the market for the latest sophisticated equipment, the farm companies will feel the loss directly and quickly.

During the 1960s, much of the management effort at Massey had been put into offsetting this, and trying to smooth out the path that sales and earnings might take in the event of a farm recession. Thornbrough had spread the company determinedly into a number of markets. If its sales were widespread, then a downturn in one market, in South Africa or Britain or France or Australia, could be balanced out. Providing the difficulties were local ones, caused, for example, by governments imposing credit restrictions, or regional food surpluses, the impact would not be so serious. And Massey had spread itself very wide indeed. By the mid-1960s, it had thirty-six plants in 10 countries, and was selling its products in 165 countries and territories.

In an effort to build up its sales in the U.S., it had set up a new headquarters and plant in Des Moines, Iowa, in the heart of the corn belt.* It had expanded farm machinery and diesel engine production in Britain, and enlarged its tractor plant at Beauvais in France. In other countries, it had initiated new ventures. In Australia, it had developed the world's first self-propelled sugar cane harvester. In Spain, it had bought into Motor Iberica S.A., to create the largest tractor manufacturing

* The move into the U.S. reversed an earlier policy of withdrawal, and of considering the corn belt only as a part of the company's worldwide operations. Surveys had shown that Massey was not doing well in the Mid-west because it lacked any real presence there. The decision was then taken to buy a plant and relocate North American operations in Des Moines even though it brought criticism in Canada that Massey was pulling out of its home base. Despite the move, sales continued to lag in the U.S. and the company was slow in moving into the large-tractor market.

firm in the country. It assembled farm machinery in Pakistan,
Morocco, and Turkey, and it went into Brazil, at the invitation
of the government, where it suffered years of accumulated
losses but stuck it out in the belief that at some stage the
inflated Brazilian economy would stabilize and farm sales
would take off.

The shift into diverse overseas locations was matched by
a shift in products. Always the aim was to get away from an
absolute and total reliance on the cyclical farm business. Some
of the new products were of small importance, for example,
Massey made snowmobiles and garden tractors, but hopes were
held out for the move into industrial and construction machinery
(ICM), a market which in world terms was as large as that for
farm equipment.

The move into industrial and construction machinery, and
into diesel engines, had changed Massey. A new triumvirate of
group vice-presidents had been installed; John Staiger taking
over farm machinery, John Mitchell heading the ICM operations,
and Sir Monty Prichard in Britain running the engine products
from his base at Perkins. Still, there were limits to the degree of
diversification that could be achieved; four-fifths of Massey's
turnover was still in farm equipment. The company remained
concentrated in a cyclical business, and it remained vulnerable
to the sudden swings that had occurred in the past and would
recur again.

Thornbrough's inclination was to expand, and to keep on
expanding. As an Amercian, he had his sights set on beating the
competition south of the border. With its far-flung interests,
there was no way that Massey could operate as profitably as,
for example, John Deere, but it could — and did (by the end of
the 1960s) — come close in sales. To Thornbrough, this was a
vindication of the global strategy he had been developing for
ten years. Massey had come out of the pack and was the leading
company in tractors and combines outside North America, and
occupied third place within the U.S. Since the trend was towards
global production, the future seemed to be more promising
for Massey than for its rivals.

Problems were, however, developing. And these were to
strike at the heart of Thornbrough's strategy. For while the
company had the advantage of operating in many different
countries and of integrated production, it was also at the mercy
of a global farm recession.

In the early 1960s, world food needs and the consumption
of grain were running ahead of production. In many of the poorer
countries, harvests were down from year to year. Stocks of
grain were being depleted, and crop prices were high. This
period coincided with a burst of expansion at Massey. By the

middle of the decade, the picture had changed. The impact of technology and new agricultural methods had helped developing countries to improve food production, and the so-called Green Revolution was being hailed as a breakthrough in solving the world's food problems. The Third World, with the aid of new miracle strains of rice and wheat, was moving towards self-sufficiency.

While welcome in itself, this development created an immediate squeeze in those countries in North America and Europe that had traditionally sold their surplus grain to the Third World. Crop prices fell. The demand for farm equipment declined dramatically. And governments in the U.S. and Canada were driven to persuading (and even paying) their farmers to lay fallow a portion of their land, and to produce less food.

The crisis had arrived. And as it turned out, Massey was ill-prepared for it. Sales had fallen off but, after its expansion spree, there was no way that production could be trimmed fast enough. As a result, more and more unwanted machinery piled up in the dealers' showrooms.

The overgenerous credit arrangements that prevailed in the industry began to take their toll, and the company became seriously strapped for cash. Sales were actually being reported at the wholesale level, and a notional profit figure arrived at, when the shipment to the dealer took place. This was because the industry financed its dealers until a retail sale was made. The problem was that farmers were not buying, and the retail sales were going unmade.

When the year 1970 closed, the company had $730 million in inventory and receivables, as against equity capital of $390 million. In the space of two years, its current liabilities had jumped by more than $100 million to $440 million. During this time, it had reported a steady climb in sales and earnings, but in fact many of the sales were represented by equipment that sat on the dealers' lots.

How could the company get itself out of this mess, and still look relatively good? How could it avoid having to report a huge loss, and how could it move its surplus inventory out? The solution that Thornbrough and his financial advisers came up with was clever, and a little devious; and it almost worked.

Obviously a first requirement was that production had to be cut back, and the dealers had to be given leeway to move the equipment, if necessary by slashing prices. Since the company had already recorded such equipment as a sale, this would mean taking writeoffs. But here the accountants had a way out. What could be done, they said, was to debit the already recorded "profits" to the company's earned surplus, which would mean that they would not show up in the 1970 profit and loss statement. A ruling by the American Institute of Certified Public Accountants had given an opinion that this was a legitimate action.

If Thornbrough could proceed on this basis, then his problems were small. The liquidity squeeze would be overcome, the surplus inventory would be worked down, and he could use the funds that would flow back to Toronto to pay down the company's short-term debt. Massey's earnings record would be preserved, while the losses would be put down as "extraordinary." This seemingly perfect, practical solution was put into effect. Very soon receivables were falling, and dealers' sales were running ahead of shipments from the factory, an indication of returning health.

The blow came only weeks before the end of the financial year. Massey's stock had been listed on the New York Stock Exchange in 1966. Now the Securities and Exchange Commission (SEC) in the U.S. put pressure on the accounting profession to change their recommendation; losses should not be swept away in this fashion, insisted the SEC, because it was misleading to shareholders. The reversal cost Massey, and Thornbrough, dearly. The company's accounts had to be altered, and it was forced to report its largest ever loss, $19.7 million.

Other factors were also at work. Production in Britain had been hit by wildcat strikes and there had been a worldwide fall in demand for combine harvesters. But the main brunt of the loss came from the pileup of inventories in dealers' hands, and the practice of recording these as sales at the wholesale level. Not only had the adjustment turned the company from a $30 million profit in the previous year to a massive loss, but it forced Thornbrough to suspend dividend payments — a situation hardly calculated to please Taylor and McDougald at Argus.

At the annual meeting in Toronto a few weeks later, Thornbrough was questioned by a man who described himself as having a triple interest in Massey. He was a shareholder, a long time Massey employee now on a company pension, and a strong union man. The exchange went like this.

Shareholder: Are we in the same position as Rolls-Royce in Britain?

[It was verging on bankruptcy at the time.]

Thornbrough: No.

Shareholder: Then our position is not as black as it looks.

Thornbrough: It is not as black as you are painting it.

The annual meeting was attended by 200 shareholders, and it had been thought that Thornbrough would have to handle some tough questions. As it turned out, the aforementioned question was the only critical one. It was a small incident but

188   for Thornbrough it was a personal rebuff. In all seriousness, a shareholder had suggested that he and his management team — who prided themselves on building a multinational company that was admired and respected by its competitors, with sales nudging $1 billion a year — had been running the company into the ground.

Any sense of proprietorial responsibility for the fate of Massey was missing.

The elderly shareholder was not the only one to be displeased by the turn of events at Massey. Belatedly, E.P. Taylor in the Bahamas (he was still chairman of the executive committee of the board), and Bud McDougald entered the picture.* They had been convinced all along, said Taylor, that Thornbrough was acting unwisely; that his estimates of future business had been unduly optimistic, and that Massey was spending too much money and needed to cut back.

Economies had to be made, and Thornbrough was prepared to make them. Nearly 2,500 workers were laid off at plants in Toronto, Brantford, and Detroit; 3,000 staff in international operations lost their jobs; and the capital spending budget was trimmed by 20 percent. Recovery proved to be elusive. There were strikes in both U.S. and Canadian plants. Management strove desperately to prevent a second year of losses, despite the disruptions and inflationary rises in costs. When the books closed on the end of 1971, they had achieved their goal largely through rigorous cost-cutting and layoffs. Still, the results were not impressive: Massey had managed to earn $9.3 million on sales of over $1 billion.

The poor performance was reflected in the price of Massey's stock. During 1969, it had reached a high of $25.50. When the trough was reached, it was trading as low as $8.40.

For Argus, this was unsettling. Its holding of 2,850,000 shares (out of 18,195,450 outstanding) was being devalued, and Taylor who had continued to buy at the $25 level was particularly unhappy. Argus had in any case been turning in a less than buoyant performance; it had reduced its holdings to six companies — Massey, Dominion Stores, Domtar, Hollinger, Standard Broadcasting, and B.C. Forest Products — but corporate assets had scarcely grown, and income levels had been depressed.

---

* Thornbrough had informed Taylor and McDougald personally of the loss that might have to be sustained in 1970 at a meeting at the Woodbine Race Track the previous September, and had told them he intended to institute major cutbacks.

In his hideaway in the Bahamas, Taylor brooded about the turn of events. He needed more money for his own New Providence Development Company in Nassau, and he was alarmed about the state of Massey. In the fall of 1972, he wrote to McDougald suggesting that they wind up Argus and share out the proceeds. The cornerstone of this was to be Massey. Taylor argued that the company would be in trouble for some time to come, and he was concerned about Ottawa's plans to impose Canadian corporate taxes on the earnings of foreign subsidiaries of multinationals. If this was applied to Massey, it would reduce its profits substantially.

Thornbrough had been concerned about the same thing. He had made it known that, if the Canadian government went ahead and imposed these tax provisions, Massey might consider withdrawing its head office from Canada.* In advancing this viewpoint, he was speaking on behalf of the board, including Taylor.

In spite of this, Taylor continued to formulate his own proposals, which were to sell out and get out. "I suggest that we do not delay the selling of our holdings in this company if we can obtain a reasonable offer within the next ninety days," he wrote to McDougald. Action should be taken quickly, he urged, before Ottawa had the chance to act and the stock market took note of Massey's diminished prospects. By selling within a three-month period, Taylor calculated Argus would have "more than $56 million in the till."

McDougald's reply was cool. He held off making any decision. A meeting was held in the Toronto home of another Argus associate, George Black, at which Taylor argued his point with McDougald. Also present were two other directors and principals, Max Meighen and Bruce Matthews. They agreed to look into the tax implications of selling their Massey block, along with the interest in Dominion Stores. Meanwhile, McDougald through his private company, Ravelston Corporation, and in alliance with Black, Matthews, and Meighen, had moved to establish effective control over Argus and its future.† He did not want to unravel the intricate web of Argus and the power and control he now held. For this reason — if for no other —

* Thornbrough made representations on the tax question to Liberal finance minister John Turner and said that it would unjustly discriminate against an old-established Canadian multinational. Turner agreed, and the proposals were shelved.

† In terms of control of Argus, voting shareholders were the Ravelston Corporation (with 776,000 common shares (45.8 percent)), consisting of McDougald, Black, Matthews, and Canadian General and Investment Companies (Meighen); Shawinigan Industries, representing Power Corporation-Desmarais (with 175,484 common shares (10.37 percent)); and Windfield Farms Limited for E.P. Taylor (with 169,278 common shares (10 percent)). Taylor, through Windfield Farms, also owned over a million nonvoting preferred shares.

he did not intend to liquidate the stake that had been built up in Massey.

The episode was an extraordinary one. Taylor's disaffection was to lead, two years later, to a bid by Montreal financier Paul Desmarais of Power Corporation for control of Argus. Taylor encouraged Desmarais while McDougald, with some adroit manoeuvering, managed to fend him off. Nominally, the status quo therefore remained as it had been. Argus continued to be represented in the same way on the Massey board and to dominate the affairs of the company, acting supposedly in the interests of all shareholders. The only visible change was that Taylor, who had openly expressed his dissatisfaction with the way things were being run, relented and stood down as chairman of the executive committee, to be replaced by McDougald. What was still missing, however, was any sense of proprietorial responsibility for the fate of Massey.

On the basis of one year of losses, and a threatened tax change that never actually materialized, the chairman of the executive committee of the board had felt it prudent for Argus to sell its controlling interest, and to do it hurriedly in order to get a good price. Nor was his desire to sell confined to private correspondence with McDougald. Taylor took it on himself to approach prospective buyers in Britain, and even one in the Middle East. It was an extraordinary manoeuver, especially for the chairman of the executive committee, who was supposed to represent the interests of all shareholders.

The Argus position in Massey, so creative in the days of Eric Phillips, was now one of passivity and disinterest.

Things were in disarray. The executive committee, designed by Argus to be the instrument of its authority, met infrequently. Taylor was seldom present. In its absence, more power was vested in the board and Thornbrough now referred to it matters that had been the exclusive domain of the committee before. The board, however, met only quarterly. And there were limits to the volume of work, or the degree of involvement, of its members. Moreover, certain board members had conflicts of interest (One example was Charlie Gundy of Wood Gundy whose investment firm had an obvious interest in Massey's plans and financings).

The break within Argus had also exposed the limitations of a closed-end investment company. Argus was based on the idea of joint initiatives being taken by its members; if, for example, additional capital was needed, then the shareholders themselves would have to be assessed or borrowing would have to be undertaken. If any of the partners had to go to the banks for new funds this would create an interest cost. The Argus members, divided as they were among themselves, were not prepared to put more capital into their own company, nor would they invest more heavily in companies like Massey that they controlled.

The result was a stalemate. McDougald did not want to confront Taylor. He often said that he had known Eddie for a long time, and that he would not be the one to force him out of Argus. This situation continued even after Taylor had broken ranks and given his approval for Paul Desmarais to mount his abortive takeover bid. Other Argus members had lined up behind McDougald; the partners of Ravelston (McDougald's holding company which had now come to own about 60 percent of Argus common stock), Max Meighen, Bruce Matthews, and George Black. But this did not clear up the situation, or help to solve Massey's problems. The company had grown rapidly. Its sales were three times as large as they had been in 1956, but during this period, with the exception of $100 million in preferred shares, there had been virtually no infusion of new equity. As a result Massey's financial position had begun to deteriorate.

Management was not blind to this. In the mid-1970s, when Massey's stock was trading at a premium, Thornbrough repeatedly raised with McDougald the possibility of a rights issue and raising new equity. McDougald was defensive. A new share issue would have meant a dilution of the Argus position unless he was prepared to put up new money himself. And after his split with Taylor, McDougald was engaged in a holding action; he wished to preserve the fabric of Argus without taking any risks or making new commitments.

He was also concerned about the opportunity that might be opened up for an unfriendly takeover of Massey. A dilution of the Argus stake would increase the danger. At the time Chrysler had shown interest, and had considered buying a majority share of Perkins though it later decided against it. To guard against takeover attempts from outside, Argus had appointed the senior executives of some of the companies it controlled, including Thornbrough, on to its own board. This move created a conflict for these executives. It was more difficult for Thornbrough to press for an additional strengthening of Massey's financial base from inside the Argus boardroom.

Intimations of trouble.

As the 1970s unfolded, it became increasingly clear that Massey was operating under the guidance of Albert Thornbrough, and decision making at the top had become increasingly subjective. Under the personal direction of Thornbrough Massey had the capability to be extraordinarily successful or to fail spectacularly. The mood of the company was different from the earlier days. It had, for all practical purposes, been set adrift by Argus. McDougald would offer his advice and sometimes get more actively involved; when a decision was made to acquire the White Motor Corporation diesel plant in Canton, Ohio, it was his task to get the best possible deal out of S.E. "Bunkie" Knudsen at White. But in general McDougald's involvement was not an active one. He took pride in Massey's achievements, but he was plainly not going to be responsible for its failures, or its future.

By the mid-1970s, Thornbrough had been at the top for two decades. He was nearing retirement age and, while his commitment remained as strong as ever, his position had become an isolated one. Massey had grown phenomenally, and he had presided over that growth. In the banner year of 1976, sales of farm machinery alone were worth more than $2 billion and total sales came in at $2.8 billion, while earnings were a record $118 million. Massey had interwoven itself into the world's farm economy, and seemed to have succeeded in attaining almost all of its goals. But success had also led to a sense of detachment on the part of management, and of isolation, a willingness to sanction new expenditures and take on new commitments with almost cavalier abandon.

Thornbrough had some able and tough executives working under him, but the chain of command as it existed and developed after 1971 did not work well. There were internal rivalries and clashes. And there were difficulties in the relationship between Thornbrough and his senior staff; as far as they were concerned, he had become remote and unreachable. Having achieved a great deal, the president of Massey had become an unshakable optimist. To inform Thornbrough of problems was to run the risk of being considered a defeatist and a negative thinker.

The atmosphere at company headquarters was still work-manlike. And the company was to score some successes; in 1976, without seeming to have overextended itself, its stock was, for a time, the favorite recommendation of nearly every financial analyst on Wall Street. Massey's international operations seemed to have put it out ahead and it was growing while International Harvester and Deere were running into problems.

Beneath the surface, however, there was an uneven quality about the management of Massey, and about Thornbrough's ability to fly solo. More ominously for the future, there was a startling contrast between management's view of things and what was actually to occur.

The crisis of 1970-71 carried lessons for the president. It

had shown how vulnerable Massey was to troubles that might
crop up in one or two major markets, and how rapidly it could
be overtaken by a serious liquidity squeeze. Thornbrough
seemed to tuck these lessons away and forget about them. In the
next few years, Massey was to run tremendous risks, out-
spending all its competitors and financing the expansion
exclusively through debt issues and new bank borrowings.
Highly leveraged as it was, the company's return on equity
looked splendid for a time. But when the farm cycle took a
downturn and earnings plunged, the whole edifice came crashing
down. The company was burdened with a staggering load of
interest and principal payments that it could not meet.

Thornbrough himself was prudent and conservative by
temperament. Yet he made decisions on capital spending and
expansion that were extraordinarily risky. He was a chief
executive who was acknowledged by his subordinates to be
considerate and attentive to their point of view. Yet they found
there were certain subjects, and company problems, that they
could not broach with him. The managerial system that had
been established was a responsive one in theory, and provided
for a flow of information to Toronto. Yet there was a growing
frustration about the decisions that were taken.

When things began to go wrong, Thornbrough was an
obvious target. Drawing a salary of $327,000, with a hefty
retirement income to follow, he had since the 1960s made his
home in Florida, keeping only an apartment in Toronto. He
was still an American citizen; indeed, in his last years as
president, he did not have landed immigrant status in Canada
on the basis that he was not in the country for more than half
the year.* In the early days, under Colonel Phillips, the policy
had been to employ qualified people irrespective of their
nationality, and no requirement had been laid down for
Thornbrough to take out Canadian citizenship. By the 1970s,
however, the issue of a non-Canadian running Massey-Ferguson
was a more contentious one.

In his job, Thornbrough travelled extensively. He often
stayed at his home at Boca Raton when he was meeting with
McDougald in Palm Beach, and he spent long weekends there
when he was going on trips, using Miami rather than Toronto
as a base when flying to or from Europe.† Certainly, Thorn-

---

* Thornbrough shared his nonresident status with another American on Massey's
staff, senior vice-president John Staiger.

† Thornbrough himself explains that he never took a vacation and therefore felt entitled
to add a Friday or Monday to his weekend when he had the opportunity of being in
Boca Raton. He also says that, from 1961 when he bought his home until his retirement, he never spent as much as fourteen consecutive days in Florida. He dismisses
criticism of his frequent flights such as "a cheap shot."

brough was dedicated to his job. After he ceased to be president he resolutely kept up his links with the company, even when he was being criticized for his part in creating Massey's problems, doing so out of a sense of responsibility. But he did not improve his image with his preference for Florida, particularly when things began to turn sour.

The first intimations of trouble came in the fall of 1977. There had been a lengthy strike at the Banner Lane tractor plant in England. Because of the strike, dealers around the world had been clamoring for additional shipments and the company had been rushing to fill the extra orders. But in September, when projected figures were drawn up for 1978 sales, they were substantially down. The decrease was not due to production problems. There had been a dramatic decline in new orders. Because of this, the Banner Lane plant would have to operate at well below capacity. Over the next few weeks, the implications of this became clearer. Management at Massey was going to have to be prepared for a major downturn the following year.

Thornbrough was already concerned about overstaffing. A study done six months earlier had shown that, for the same level of operations, Deere employed 5,000 fewer people than Massey. Previous instructions to cut back staff, particularly in Europe, had gone unheeded. Now Thornbrough ordered a further retrenchment. He also ordered cutbacks in production. Aware of the impending recession, he was forced to announce that dividends on both preferred and common shares would be suspended until further notice.

Massey was plainly on the verge of a major setback; when figures for the first quarter were announced, they revealed a loss of $38.7 million, far higher than anyone outside the company had expected. The cancellation of dividend payments had also been made necessary by some tough conditions attached to a $300 million U.S. loan. Taking note of Massey's debt position, American institutional lenders had insisted that dividends must not exceed $30 million plus 75 percent of consolidated net income after November 1, 1977, or minus 100 percent in the case of a loss. Despite the restrictions, Massey had been forced to go ahead with the borrowing. It needed the money desperately to repay current bank loans and handle debt maturities of $95 million.

The terms of the private financing deal indicated just how serious Massey's problems were. The company had been operating on the basis that recessions were a thing of the past, and that it could afford to expand indefinitely.

What had been occurring, in fact, was a wholly adverse trend in the world economy, which Thornbrough and his senior

executives had chosen to ignore. When markets turned down,
Massey found itself with too much capacity and too many
commitments. Worse, it was saddled with servicing the huge
debt it had taken on during its years of expansion.

Some of the obligations that Massey had incurred seemed
to be justified at the time. After all, management could not
have been expected to anticipate the oil crisis of 1973 and the
problems it would bring, and expansion abroad had seemed
very feasible at the time.

In Brazil, Massey had hung on through some grim years
and had built six plants. By the mid-1970s, this patience seemed
to be paying off; the company had 50 percent of the tractor
market, and 20 percent of all industrial and construction
machinery sales in a fast-growing market. Sales in Brazil were
second only to the U.S. A short time later, the Brazilian government, worried about oil deficits and galloping inflation, clamped
down on lines of credit to farmers. Suddenly, the bottom fell
out of Massey's market in Brazil. In Turkey, the company had
carefully built up its position. As a result of its long time
association with the Turkish government, Massey had captured
half the tractor market. Looking ahead, it was not difficult to
envisage the country providing a market for 60,000 tractors a
year. After the oil crisis, the outlook in Turkey changed entirely.
By 1978, the Turkish economy was on the verge of bankruptcy.

These were instances of bad luck more than bad judgment.
Looking at the situation in the early 1970s, the problems that
were later to occur were difficult to forecast. What could have
been done, however, was to minimize the risks. For every
example of the company being beset by misfortune, and by
events beyond its control, there was an example of plain errors
of judgment. Management at Massey had been too optimistic
and trusting.

A great deal of money was put into developing a line of
large tractors in the U.S. But the big tractors were plagued
with production-line difficulties and were finally launched on
an unreceptive market. In response to pressure from Monty
Prichard of Perkins who was eager to get into the large-diesel-
engine business, competing with giants like General Motors and
Cummins Engines, a deal was struck to take over the plant and
engine design of White Motor Corporation in Canton, Ohio.
Later, Perkins engineers announced they could not convert the
White engine, and the product line had to be scrapped. The
Canton plant was switched to four- and six-cylinder engines,
and its output was a meagre 22,000 units out of a capacity of
75,000. Poorly conceived and badly managed, the projected
expansion of Massey's diesel engine company into the U.S.
turned into a dismal failure.

The biggest embarrassment of all, however, was the purchase of Hanomag from a West German steelmaker, Rheinstahl AG, in 1974. For $45 million, Massey got itself a major construction machinery firm based in Hanover, with a three million square foot plant, modern machine tools, and a product line of wheel loaders, crawler tractors, bulldozers, and hydraulic excavators. The factory was to be used to produce a new range of Perkins engines, and its purchase (according to the company's press release) represented "a major strategic step in the expansion of Massey-Ferguson's construction business."

Hanomag proved to be a disaster. It had been bought at the wrong time, just as four boom years for construction machinery came to an end in West Germany and the deutsche mark began to strengthen against other world currencies, holding down exports. Massey's attempts to sell high-cost Hanomag machinery through its dealer network were less than unsuccessful and the losses mounted up.

Not unnaturally, these blows hit hard back in Toronto. The strategy that Thornbrough had shaped was coming apart, and the losses to Massey's prestige as well as its balance sheet were staggering. Thornbrough was not given to recriminations, nor was he going to walk away from his post, despite the fact that he had already attained retirement age and had become deputy chairman, a promotion that would have made it easier for him to turn over executive responsibilities to someone else. Indeed, to many, he seemed altogether too reluctant to stand down.

Part Four:
At the Brink (1978-1981)

12

The Barons Do Battle

But when, in early 1973, a new generation of ambitious executive vice-presidents took to the field, the rules were altered and the game suddenly became a rough one.

Five days before Christmas 1972, Massey-Ferguson announced that its senior management positions were being swapped around. The news was not greeted with much fanfare. In Canada, Massey did not often make the headlines. Although the firm had offices in the heart of Toronto's business district and its factories and farm equipment dealerships ranged across the country, Massey had deliberately de-emphasized its Canadian identity.

Most recently, Massey had made the headlines by threatening to move its head office out of Canada. The Trudeau government in Ottawa had been considering toughening the rules on the taxation of foreign subsidiaries. Most Canadian companies could not have cared less, but Massey would have been hurt badly. Thornbrough had protested both publicly and privately. And the proposed new rules had been dropped. After this brief flurry of publicity, the company and Thornbrough retreated again from public view. So, when it was announced that Massey was getting "a management structure that will enable it to respond to the opportunities of today's environment and those of the foreseeable future," to quote a Massey press release, the news generated no excitement and little coverage.

Massey had long been a company that believed in shifting operating responsibility into the many countries where it did business. Its organization gave a great deal of authority to the managers who knew the market conditions in Brazil or South Africa, Australia or Argentina. This autonomy, it was felt, enabled the proper decisions to be made as to which products

198 should be built and where they should be marketed. And the strategy had worked: Massey-Ferguson was very successful, and very big. Its sales in 1971 had topped $1 billion, putting it among a select group of only two or three Canadian companies.

The diverse nature of Massey was reflected in the senior management changes that were announced. Of the ten executives who were to be given the senior jobs, four were Canadian, four were American, and two were British.* Under the new arrangement, the most powerful positions were to go to executives who would rule over the foreign operations of Massey.

Massey had reorganized itself before. It had changed its top team in the late 1950s after James Duncan had been ousted. It had revamped its senior positions again in 1968, as the spread of its business became larger and more decentralized. Now it was making a third change in a conscious attempt to keep pace with its own growth and the complexity of its operations around the world.

In the scheme of things at Massey there were two people who mattered. One was Thornbrough, now sixty years old and in charge of the company as its president for the past sixteen years. The other was Bud McDougald, the chief of Argus Corporation. McDougald exercised his ruling power through his chairmanship of the executive committee of the Massey board, and made his views known to a compliant board of directors, consisting of his friends from Ravelston and Argus. They could be counted on to do his bidding and listen to his often-extreme opinions on a variety of subjects. Moreover, he had formed an alliance with Thornbrough, confirmed by the election of Thornbrough to the board of Argus.

The changes taking place at Massey were McDougald's idea. He had not been greatly pleased by a sudden dip in the company's record of profitability in 1970 and 1971. The sudden slump did not unnerve Thornbrough. But McDougald began to exert pressure on him to think about a successor. The company had become exceedingly large and unwieldy and there were many executives reporting to the president. While this might be manageable for the present, McDougald was concerned that

* The ten senior appointments which created new posts and confirmed old ones were: John Belford, vice-president, Personnel and Industrial Relations; Ralph Bibow, vice-president, Farm Machinery; Peter Breyfogle, vice-president, Corporate Operations; Dave Goodson, vice-president, Industrial and Construction Machinery; Wally Main, vice-president, Administration; John Mitchell, executive vice-president, Americas; Sir Montague Prichard, executive vice-president, Engines Group; John Staiger, senior vice-president; Hugo Vajk, executive vice-president, Asia, Africa, and Australasia; Peter Wright, executive vice-president, Europe. Of the four Canadians, Belford, Main, Vajk, and Breyfogle, two were naturalized citizens; Vajk had been born in Yugoslavia, Breyfogle in Spain. Of the remaining six, Prichard and Wright were Britons and the other four Americans.

further growth would put strains on Thornbrough, and on his
ability to cope.

Thornbrough's response was to create an organization and management committee, made up of long-serving, senior executives loyal to him, and seek their advice on how he should proceed. Thornbrough and his advisers were men who had had long experience and were in no hurry to step down or surrender the positions they had acquired; nonetheless, they had to acknowledge that there was a real problem at Massey, and that a company which was forecasting worldwide sales of $2 billion or more in the early 1970s could not continue to depend on a president in Toronto for so much of its decision making. More effort had to be made to decentralize, to create other centres of power. And, in the process, the executives to whom power was handed would come to be thought of as potential successors to the president.

The recommendation of Thornbrough's committee was approved by McDougald. It was this radical change that was contained in the announcement of December, 1972. Henceforth Canada's oldest and largest manufacturing company would become more international. A large measure of authority would be exercised by four newly created executive vice-presidents, and the power they were to be given would be exercised outside Toronto — often outside the purview of Thornbrough himself.

There was to be an executive vice-president in charge of the Americas (North and South); another would take over Europe; and a third the markets of Asia, Africa, and Australasia. The fourth would be in charge of Perkins Engines in Peterborough, U.K. a business Massey had always treated separately.

The dilemma Massey faced was not an unusual one. In the history of other corporations similar difficulties had been encountered, and a similar solution arrived at. In effect, mini-presidents were created to run divisions of larger companies. And their success or failure in doing so had determined their fitness for the chief executive's chair. The experiment at Massey, however, was to be disastrous. Perhaps more than any other single element it was to lead to the ruination and near-bankruptcy of the company a few years later. The ultimate responsibility for what was to befall Massey-Ferguson rested with Thornbrough and McDougald. But neither of them could have known how destructive their decision to reorganize management and open the way for a new president was to prove.

There are two views of Albert Thornbrough in his last years as president of Massey-Ferguson. The first is held by his close associates and old friends who saw nothing machiavellian in his nature, and contend that he was genuinely interested in a new, younger man coming forward and taking the presidency. The second opinion holds that Thornbrough went through the exercise of creating a quartet of executive vice-presidents under pressure from McDougald but was really unwilling to step down or surrender his position at any time. The failure of any of the new contenders to achieve the presidency is seen as proof of this. Thornbrough did not want to give up and held on to the last.* Evidence can be put forward to support both sides, but none of it is conclusive (Thornbrough's motives remain his own). But what is not in question, is the lack of leadership and strong direction that became apparent at Massey. Never decisive and seldom ruthless, Thornbrough had built a company that reflected his personal traits; he was competent, hardworking, and capable, but also dangerously weak and irresolute, liable to be torn apart by argument and conflict.

Up until now, management at Massey had avoided dissension. The views that prevailed at Massey were those of Thornbrough. But when, in early 1973, a new generation of ambitious executive vice-presidents took to the field, the rules changed and the game suddenly became a rough one. In the four years that followed, Thornbrough was to show himself incapable of acting as a referee.

The first executive vice-presidents to be appointed were all long-serving company men; an American, John Mitchell, was given control of North and South America (the Americas); a Briton, Peter Wright, was given Europe; Hugo Vajk got Asia, Africa, and Australia or, as it was known within Massey, the three A's; and another Briton, Sir Monty Prichard got Perkins. A key position at head office went to Peter Breyfogle. He was made vice-president, Corporate Operations, in charge of all head office activities. In the next few years this chief-of-staff post was to become a crucial one. When Breyfogle was promoted to executive vice-president in Europe in 1975, the young British

---

* An earlier contender as Thornbrough's successor had been a Polish immigrant to Britain who had joined the Ferguson Company as a shipper after the war. John J. Chluski rose swiftly through the company ranks in the 1960s, becoming vice-president in charge of North American Operations. In 1968 he suddenly and unexpectedly resigned from this post, feeling that Thornbrough was not prepared to back him in making the necessary outlays that would have improved Massey's position in the U.S. Chluski had been the first who, seemingly, was being groomed for the presidency.

deputy managing director of Perkins, Victor Rice, was shunted
— much against his will — to Toronto. Rice was to build up a
power base within the company, first as comptroller, then as
vice-president, Corporate Operations, and eventually push his
claims for the presidency.

The emergence of Rice was, however, a long way off. And
it seemed in the early 1970s a remote possibility. Instead the
attention was focused on the four men who presided over the
geographical spread of Massey's interests. One of them,
it seemed, would likely be Thornbrough's successor and since
the firm had always demonstrated a remarkable staying power
in keeping and holding its presidents (Thornbrough and his
predecessor, James Duncan, each presided over the company
for twenty-one years), the prize appeared to be one worth
fighting for.

The contenders were a disparate group. Critics of Thornbrough later were to claim that this was intentional, a part of
his scheme for maintaining his own hold on power. If the
president had really committed himself to the notion of someone
succeeding him, then why did he select executives whose temperament and style made them less than suitable for the job? Why
did he persistently try and play one off against another? Often,
posts were filled with executives who could be counted on
to oppose each other, thereby negating each other's influence.
Moreover, Thornbrough's control over the company was
erratic and unreliable; increasingly he was seen to be playing
the role of some medieval king trying — and failing — to keep
the power of his barons in check.

The first of the powerholders to count himself out was Sir
Monty Prichard. At Perkins, he had long ago created his own
enclave. At the time of the acquisition by Massey, he had also
managed to get a commitment from Massey that, because of
Perkins' unique position as a supplier to other clients, it should
be left to run its own affairs and allowed a greater degree of
autonomy than other Massey subsidiaries. Prichard was an
aggressive and forthright executive, a hard-talking salesman, and
he wielded the document that gave him power over Perkins
like a weapon. The arrangement was not a satisfactory one for
head office in Toronto; the company relied on Perkins as
much as it did on its other manufacturing plants and, since
production around the world was becoming closely integrated,
the waywardness of Perkins — and the wall that Prichard had
built up around the company — became an increasing source of
irritation.

Prichard had great plans for his firm. He wanted more
money for expansion. He saw opportunities for Perkins to move
into the production of larger diesel engines, and he submitted

one report after another on the need to build a new plant in continental Europe, or acquire a new factory elsewhere. There were some merits in his plans. But Prichard did not stay with the firm.

He felt he had been frustrated in his plans for Perkins. Thornbrough often seemed to be encouraging him, safe in the knowledge that McDougald, who held the purse strings, would not approve expensive new ventures and they would ultimately be vetoed. In the end, Prichard wrote Thornbrough a letter detailing his personal plans; either he would go off and found a think tank in the Caribbean, or he would create a new joint-venture company with Perkins, or he would take early retirement. Thornbrough wrote back that he would accept the third alternative.

The departure of Prichard caused a reshuffle in Massey's senior ranks, and brought to the fore the lineup of executive vice-presidents who were to contest strongly for the president's chair.

Peter Wright, the Englishman who had headed the company's export operations, had initially been put in charge of Massey's interests in Europe. He began to construct a corporate organization which did not meet with the approval of all the national interests Massey had to satisfy; the French complained that Wright and his staff were Anglo-Saxons and did not have a proper understanding of their needs. Since France ranked as the second biggest market in Europe, the criticism registered at head office. Wright was transferred to Perkins while Peter Breyfogle was sent to Britain as the new executive vice-president for Europe. A capable executive, Wright had no burning ambition to reach the top. He was determined to bow out of the business world at the age of fifty, and for years had lived frugally with the idea of taking early retirement. At the end of 1977, he took himself out of the running and quit the company.

The elimination of Prichard and Wright revealed the shortcomings of the strategy being pursued at Massey. It was unlikely that either would have become president; Prichard because he did not have the confidence of Thornbrough, and Wright because he had no desire for the job. But their departure clarified the situation and the lines of battle that were emerging among the heirs apparent. It was becoming increasingly obvious that there were three choices for Thornbrough's job, Mitchell, Breyfogle, and Vajk, and the rivalry between them intensified. For each of the three, the main objective became to press his own candidacy. To do so required gaining the president's ear to tell Thornbrough how well things were being run in the candidate's specific region. Thornbrough played to this. Increasingly isolated and detached from the operations he was responsible for, he had become an incurable optimist and wanted to hear good news.

Of the three frontrunners, John Mitchell and Peter Breyfogle quickly became the favorites. Hugo Vajk was a tactician and he was able to score points with his diplomatic handling of relations with the many governments that Massey had to deal with in the Third World, but his territory in the three A's was removed from the mainstream of Massey-Ferguson, and concentrated on marketing rather than manufacturing. Moreover, Thornbrough seemed to have decided early on that Vajk did not have the determination or drive needed to run a multibillion-dollar corporation. This decision did not of course satisfy Vajk or restrain his ambition.* By the mid 1970s sales to the countries of the Third World were propping up Massey, as markets took a dramatic downturn in North America and Europe. Vajk felt that his success warranted greater recognition, and he kept on pressing his claims.

John Mitchell, who ruled over his barony which included all of North America, together with Mexico, Brazil and Argentina, from his own head office in Des Moines, Iowa, had plenty of drive. Mitchell had joined the company in the 1960s from Clark Equipment to develop its sales of industrial and construction machinery. He was a persuasive and spellbinding salesman who had risen rapidly despite the comparatively poor performance of the industrial and construction side of Massey's business. Mitchell was not much liked in the company. The Europeans had a distinct distaste for his all-American ways and his overbearing manner. He constantly and vociferously complained about the lifestyle, poor service, and general inefficiency of Europe and other foreign lands, and seemed to regard the U.S. as the only acceptable place in which to live and work. In a fit of pique, Mitchell once telephoned the president of Hilton hotels to complain about the service he was getting in their hotel in Sao Paulo, Brazil. Apparently no one in the hotel knew how to press his pants properly.

Thornbrough liked John Mitchell and listened to him. He was aware that Mitchell had many shortcomings as a potential president, but he responded to his hard-selling voluble manner and did nothing to hold him back. Mitchell had grandiose schemes. He took himself out of Iowa and Middle America to the balmier climate of Fort Lauderdale, establishing an export office for the Americas in Florida, and buying a Lockheed Jet Star which, he explained, was the only corporate jet that would enable him to commute to Des Moines and back without re-

* Vajk had been educated in France (Grenoble University), Canada (McGill University), and the U.S. (Carnegie-Mellon University), before working in Canada and for a Pittsburgh-based firm, Joy Manufacturing Company, in Paris. He had joined Massey in France in 1964 and had taken over its French operations one year later. In 1969, he had been appointed vice-president, Logistics, at head office.

204 fuelling. He enlarged his staff, and convinced both himself and Thornbrough that Massey could win a larger share of the U.S. farm market almost by default. In Mitchell's eyes, John Deere and International Harvester were on the run and Massey was on the verge of overtaking its rivals.

Mitchell was ambitious for the top job. And so was a Canadian, Peter Breyfogle. Educated at Cambridge University and the Harvard Business School, Breyfogle was forty years old when he became Massey's executive vice-president for Europe. He had joined the company as a financial analyst in 1959 and risen through the Export Organization and North American Operations, becoming comptroller and then vice-president of Corporate Operations. Bright and aggressive, Breyfogle represented a new, younger generation of Massey executives. His manner was also abrasive; he had the reputation of being a man who was impatient and in a hurry. There was no doubt that Breyfogle's appointment to Europe gave him the inside track for the presidency. And that was almost certainly the view of the senior managers surrounding Thornbrough, and probably of Thornbrough himself. John Mitchell was a powerful personality, and he could be relied on to get things done, but Breyfogle possessed all-round capabilities, including financial skills, and he was young and in good health (which Mitchell was not).

The brief Thornbrough gave his new European chief was to reduce staff, and to cut costs. Several reports showed Massey was grossly overstaffed compared with other farm equipment makers, and a lot of this excess was to be found in its European operations. By tradition, the company had grown and acquired its multinational status in Europe, and in Britain. In its best years, Massey had been able to achieve reasonable profitability by making use of lower European labor costs and selling into the expensive North American market. Things had not always worked smoothly. The British factories had a history of fractious labor disputes and strikes. And Thornbrough was concerned that costs should be trimmed, and efficiency improved.

Breyfogle was by nature an expansionist. Combining this inclination with his rising stature in the company, he was determined to win out over his rivals by demonstrating that the European operation could make great progress, and that growth was possible even under adverse circumstances. In Toronto and in Europe, Breyfogle was the champion of two major decisions; first, to acquire and convert the White Motor plant in Canton, Ohio, to manufacture a new family of Perkins engines, and second, to purchase the Hanomag construction machinery plant in Hanover, West Germany. Both decisions were much debated in Toronto before the go-ahead was finally given. Both, however, turned out to be colossal mistakes.

In Europe, Breyfogle continued to present plans for further expansion. Like Mitchell, he built up his own corporate staff and his own authority. He believed that a sure way of implementing his plans and winning over Thornbrough to his point of view was, first, to construct a head office staff of his own who could ensure that he was well briefed and his arguments would prevail and, second, to limit the amount of access that staff from Toronto could have to Europe. Under the baronial system, corporate staff had to ask permission before visiting or reporting on what was happening in the regions; and none of the corporate staff was given the power and prestige that had been vested in the regional executive vice-presidents. When Breyfogle, or Mitchell, or Vajk travelled to Toronto, they would be accompanied by a team of advisers to help put across their case. In Europe the corporate staff had grown so large that at one point Breyfogle proposed that a country mansion near Coventry, Stoneleigh Abbey, the ancestral home of Lord Leigh, be purchased to house them.

A mountain of debt.

Despite the financial problems that had been encountered in the early 1970s, Massey had been expanding rapidly. There were few markets where the company did not envisage a great future. The strategy had been to establish a beachhead in unlikely places either in the Third World or even behind the Iron Curtain, to seek out government approval and then set up a marketing or manufacturing subsidiary, and to match this expansion by increasing production in Massey's main plants in Europe and North America. By 1976, the company had 68,000 employees around the world, sales of $2.8 billion, and record profits of $118 million. It had also acquired a mountain of debt. The situation was not alarming as long as markets kept expanding. But there was no doubt that the company was highly leveraged. In 1975, an internal report circulated by John Deere to its senior management had drawn attention to the financial state of Massey, and had concluded the company was highly vulnerable if world markets turned down.

Problems were developing. The global oil crisis had introduced a new element of instability into the world economy; moreover, it had been injurious to the poor, Third World nations where Massey did much of its business. Inflation was creating difficulties for all countries. Currencies had been set free to float against each other.

Thornbrough did not see these problems. Nor did he prepare for the inevitable impact they would have on the farm equipment industry. In the monthly meetings with his senior staff and executive vice-presidents, Thornbrough received positive reports and forecasts. The emphasis was almost always on expansion and growth, and on seeking out new opportunities for the future.

At one meeting, held in Florida, a cautionary note had been struck by the comptroller, Victor Rice. He had pointed out that the results in Massey's record year of 1976 had actually been poor. At a time when markets were expanding rapidly, the company had managed to achieve a return on sales of just 4.2 percent. In addition, Rice had produced figures that showed that $50 million of $118 million in profits made that year were attributable to the extraordinary battering that the British pound had taken during the year. This had lowered Massey's costs and raised its earnings. But such an event was unlikely to occur again. What if the pound became stronger against other currencies instead of weakening? Thornbrough was exasperated by this line of reasoning. It had been a record year for the company. Bud McDougald was pleased,* and the staff would be congratulated.

Rice pursued the matter in the months that followed. He was supported by a few members of the corporate staff who were worried about Massey's overexpansion, and Thornbrough's unwillingness to recognize that the company's growing debt load could create problems. At the monthly co-ordinating meetings, where only senior executives were present, the executive vice-presidents were strident and sometimes abusive. Rice did not know what was happening in their regions and markets. He was a "numbers" man and a "bean counter" who knew nothing of the real world.

The meetings were lengthy and rambling. Thornbrough exercised little restraint over his senior staff. When they were argumentative, it would often be because Rice was being awkward and insistent that the outlook was not so rosy. A battle raged over the numbers he kept quoting to show that Massey was seriously overextended. He concentrated his fire on the construction machinery business, a favorite of John Mitchell's. The firm had committed itself to buying Hanomag but it was not in a position to back up its investment with new capital. As a result, Massey could not hope to challenge the industry's megafirms: Caterpillar, Komatsu, International Harvester, and

* Addressing the annual meeting of Argus in 1976, McDougald had declared; "All our investments seem to be doing pretty well. I think it is fortunate that we have top management and every one of them seems to be doing a good job." At the time the net asset value of the class C and common shares of Argus had climbed 27 percent in a six-month period, chiefly due to the rise in Massey stock from $20 to $30.

J.I. Case. Instead it was trapped in a money-losing situation. In
Germany the deutsche mark rose and Hanomag's production
costs spiralled upwards. Losses on construction machinery were
$49 million in 1977, and $57 million in 1978.

Eventually, Rice was to win his point and it was agreed that
the construction subsidiary should be sold off. But there were
few such victories. Fully engaged in their spirited battle for the
president's job, the trio of executive vice-presidents would not
concede that Rice's figures were the right ones; in their own
estimation, business was prospering and Massey was gaining
a greater share of the market everywhere. At each monthly
meeting some twenty-five senior executives would be grouped
around the table, while outside at least as many foreign-based
staff were present, having flown in to Toronto from London, or
Lucerne, or Fort Lauderdale, to provide support for their regional
chief.

In a sense, the monthly co-ordinating meetings were only
the most visible sign of the struggle that was raging at Massey.
Held before Thornbrough who would, sooner or later, recommend a new crown prince to the board, these meetings were of
tremendous importance to the contenders.

But while the would-be presidents fought it out in Toronto,
the real damage was being done to Massey's business around the
world. Decisions were being taken that were not in the best
interests of the company. Unwise expansion plans were being
pushed through. The company was spending and borrowing
freely regardless of the consequences. Under the executive vice-
president system and the permissive rule of Argus and Thornbrough, shareholders' equity increased a mere 15 percent (from
$470 million to $541 million) between 1973 and 1978, while the
total debt soared by 157 percent (from $779 million to $2 billion.)
By any standards it was an unhappy performance.

Within the company, there were few people in positions
of power — and none in positions of influence — to alter the
course Massey was on. Executives were divided into different
camps. Those who were part of the head office staff, and who
could be expected to have the best overall view of what was
happening, were least able to change things. One Massey
official who viewed events from London commented that in
the 1950s and 1960s, the company had talked seriously about
building better tractors and combine harvesters. By the middle
of the 1970s, all the talk was about company politics.

In this atmosphere, the business was being run down.
Development of new products and relations with dealers were
neglected. Much of the corporate effort went into proving that
one region had done well and another badly. When a competitive bid was made to the Egyptian government for a shipment

of tractors, staff in both the three A's region and the Americas region had agreed on a price. When it was found that Massey could have bid $300 or $400 more for each tractor and still obtained the order, Vajk's and Mitchell's staff were each given the job of producing detailed reports to show that it was the other regional headquarters that had made the costly final decision.

In Sri Lanka, a dealer placed a sizable order for 2,500 tractors, which were to be used for transportation and for pulling trailers. The order came with a stipulation that deliveries should be made at the rate of 500 a month because the distributor could only handle the importation requirements, and the financing, on this basis. The Banner Lane tractor plant in England was attempting to push sales to make its quarterly results look good, so the decision was made to ship all the tractors at one time and to send them off even though they were missing a vital component, an automatic trailer hitch, which rendered them useless to the Sri Lankans. Eventually, the company wound up having to dispatch the missing parts to their destination by air freight, and help out the importing and financing costs — the total bill came to £38,000.

The same reckless decision making was going on in Massey's larger markets. The company was a homogeneous one, so that each entity and each region was dependent on the co-operation and sense of responsibility of the others. Somehow, this fundamental fact was being overlooked, resulting in overexpansion and waste. The marketing group in North America might decide that markets were weakening, and they could only sell so many thousand tractors in their region. This would then be overruled by a general manager who had been instructed that a certain sales target had to be met. More tractors would be shipped to the market whether they could be sold or not. The premise behind this was that Massey was a successful company and could afford it. A young manager arriving in the Toronto plant was told not to concern himself too much with inefficiencies in production and delays in deliveries and shipments; Massey was a company that could not fail to make money.

The question of the succession was growing more confused.

The record performance that Massey had turned in during 1976 buoyed Thornbrough and delighted McDougald. By inclination McDougald was an internationalist, and he enjoyed the acceptance that Massey's name carried around the world.

signed by any western company with Poland, a $360 million
deal to modernize the country's tractor and diesel engine
production, McDougald was enthusiastic and pushed the venture
along. When top-level contacts were being negotiated with
governments and business leaders, McDougald would personally
lend a hand.

At the age of sixty-eight, McDougald did not take an active
part in overseeing Massey. Nor was he often to be found in
Toronto or in the Massey offices. Thornbrough was in charge.
And McDougald was left unaware of the internal squabbles
and dissension wracking Massey. McDougald did, however,
insist that the succession problem be dealt with. And in 1977, as it
became clearer that sales were falling off and a recession was
looming ahead, McDougald returned to the subject, informing
Thornbrough that a decision had to be made.

McDougald's insistence put Thornbrough on a spot. During
that year he had passed the normal retirement age of sixty-five.
Under the rules of the company, he had received an extension
from the board to continue as president. Plainly, he did not feel
the time had come to relinquish the post. His decision was
reinforced by reports coming in from the regions. Massey's
markets were turning down and the optimistic reports that he
received from his executive vice-presidents could not conceal the
fact that the prospects for the company in 1978 were deteriorating. Thornbrough felt he had the experience to steer the firm
through the bad times that lay ahead, and he had no confidence
that a new president would manage as well.

He was also faced with the dilemma that none of the
presidential candidates had turned out to be suitable, at least in
his eyes. He had ruled out Hugo Vajk. John Mitchell and Peter
Breyfogle were still aggressively canvassing. But Mitchell was a
sick man, and Thornbrough was altogether disillusioned with
Breyfogle's performance. Looking around the senior ranks of
the firm, there seemed to be no other choices. Victor Rice was
ambitious and clever. However, to Thornbrough, he lacked both
experience in the field and maturity; he was just thirty-six
years old.

In conversations with McDougald, it was decided that
Massey should look outside for a president and attempt to
recruit an experienced executive from another major multinational company. McDougald gave his approval to this plan,*
but, shortly afterwards, in March 1978, he died at his Palm
Beach home.

The death of McDougald postponed the plan to find an

---

* On January 18, two months before his death, McDougald had finally appointed
himself chairman of the board as well as chairman of the executive committee.

210 outsider. And it launched a wider struggle for power. The leadership of Argus, and control over the board of Massey, was suddenly thrown open. Thornbrough found himself dealing with new directors at Argus who were far less receptive to his views than McDougald. He had lost an important ally. Meanwhile, within Massey itself, the question of the succession was growing more confused.

Most of the Massey senior executives felt that Breyfogle would get the call. And their opinion was reinforced when Thornbrough took the title of deputy chairman of the board; clearly a move designed to make way for a new president. Victor Rice, who had opposed Breyfogle, and clashed with him at the monthly meetings, was resigned to what would happen and thought that one of Breyfogle's first decisions would be to have him fired.

Thornbrough, however, had other plans. Breyfogle had been sent to Europe to do a particular job. He had been told that he must reduce the corporate structure, which had grown excessively, cut back levels of inventories and generally improve profitability, and that he should concentrate on new product development. In Thornbrough's opinion, none of these things had been done. While he admired Breyfogle's skills as a financial analyst, he felt that management in Europe had been poor and that the reports he was getting did not square with the facts. Moreover, the situation was becoming critical. Forecasts of sales and earnings for Europe in 1978 made dismal reading. For all these reasons, Thornbrough told the executive committee he could not recommend Breyfogle remaining in Europe any longer.

The executive committee then gave its approval for what was to occur. Breyfogle was told that Thornbrough wanted to meet with him in Toronto in April. Blissfully unaware of what was in the offing, he arrived expecting to be told that the job he had struggled for — the presidency of Massey-Ferguson — was his. Logically, he seemed the only choice. While the situation for the company was bad in Europe, it was not much better in any other region, and Breyfogle felt that his twenty years with the company, his seniority, and his impressive academic background merited the recognition he believed he was about to be given. Instead, there was a stormy meeting with Thornbrough. Breyfogle offered his resignation, and it was accepted.*

Within the company, the news of Breyfogle's departure had a stunning impact. Having been engaged for six years in skirmishing for Thornbrough's job, it now seemed likely that none of the barons would get it. Only Hugo Vajk remained as a contender. And while he was now the favorite, the probability of his

* Breyfogle subsequently went on to become vice-president, Finance, for Dome Petroleum of Calgary, one of the country's most expansion-minded companies.

succeeding quickly, or succeeding at all, seemed remote.
Thornbrough had in fact wiped the slate clean. At a particularly critical time, with Massey's fortunes waning and Argus in the throes of dealing with its own succession problems, he had eliminated the likeliest successor to himself. Far from diminishing the uncertainty about the future of the company, Thornbrough had added to it.

Part Four:
At the Brink (1978-1981)

13
The Denouement

Massey was faced with a crisis which was deepening and becoming more serious. But its needs were to be subordinated over the next few months to the drama that was unfolding at Argus and the contest that was taking place among McDougald's colleagues and presumptive heirs.

The death of Bud McDougald was a momentous event; for the Canadian business community, which had come to regard him as an almost legendary figure, for his partners in Ravelston and Argus, and for the president and now deputy chairman of Massey-Ferguson, Albert Thornbrough. McDougald's funeral was a grand occasion which brought out the elite of the business world to pay their final respects; men of influence and reputation themselves, they were not slow to recognize that with the death of McDougald an older generation of entrepreneurs had passed from the scene or stood down from positions of power. The Argus group had jealously safeguarded their interests in the companies they controlled. But when it came to clarifying the line of succession at Argus, the final arbiter had been McDougald and he had been less than clear about his intentions.

A few years earlier, speaking before a royal commission set up to investigate corporate concentration, the most powerful businessman in Canada had declared: "When I croak, my colleagues will still be around and they will find some young shareholders to carry on."*

---

* Another McDougald observation on the Argus succession was made to Hugh Anderson of the *Globe and Mail* at the time of the Desmarais takeover bid. "The succession at Argus is not a problem," McDougald said, "because there are agreements among members of the controlling group requiring their interests to be offered to the others in the event of death or departure from the group for any reason." As it turned out, there were no agreements, only disagreements.

It was an enigmatic statement since the men who regarded themselves as the natural heirs to McDougald were by no means young, and in no way disposed to surrender the positions of power they had finally attained. In the spring of 1978, they moved to consolidate their hold on the satellite companies of Argus. In doing so, they posed a new and disturbing challenge to Albert Thornbrough who seemed to be more firmly entrenched than ever at Massey. Thornbrough had always managed to establish his credentials, and build up a working partnership with the strong leaders of Argus, first Eric Phillips and then, after his death, Bud McDougald. But he was on much less firm ground with the purposeful but aged triumvirate that emerged after McDougald's death.

Maxwell Meighen, aged sixty-nine, was the second son of a former Conservative prime minister,and had been a member of the Argus board for seventeen years; Bruce Matthews, sixty-eight, a former major-general in the Canadian army, was a confidante of McDougald and had been executive vice-president at Argus; Alex Barron, at fifty-nine the youngest and most vigorous of the three, had teamed up with Meighen in an investment firm, Canadian General Investments, which held an important stake in Ravelston and therefore in Argus. All three were directors of Massey. Within a month of McDougald's death, Thornbrough was faced with a new board of directors. Matthews was elected chairman. Barron took over the key position as chairman of the executive committee of the board.

The first task both men set themselves was to bring some order to the haphazard relations that had existed between the directors and management at Massey. Directors' meetings had been casual and infrequent, often given over to long diatribes on some political happening that vexed McDougald. As a military man, Matthews ran the board in a more disciplined fashion. Barron took his role as chairman of the executive committee seriously, and began to explore what had gone wrong with the management of Massey. On several occasions, Barron spoke to senior executives rather than to Thornbrough about the problems the company faced. He also decided that Massey was being badly served by its present management, and that changes had to come swiftly.

These initiatives lasted only a few months, however. For, in a matter of weeks after McDougald's death, it became clear that a bitter struggle was developing for control of Argus. Although Massey was faced with a deepening crisis, its needs were to be subordinated over the next few months to the drama unfolding at Argus.

The balance of power at Argus was vested in the Ravelston Corporation, a company McDougald had formed in 1968. The company's only significant asset was its 62 percent stake in the voting common shares of Argus. It had been formed to hold the shares of McDougald himself, of Eric Phillips' estate, administered by McDougald, and of three associates; Meighen, Matthews, and George Black, a Winnipeg accountant and former president of Canadian Breweries. Ravelston had shown its effectiveness a few years earlier when Paul Desmarais, head of the Montreal-based Power Corporation, had launched a takeover bid for Argus. McDougald had rallied his troops, and they had agreed not to sell to Desmarais. From that point on, control of Ravelston, a private company tucked away from public view, had been the key to unlocking Argus, and establishing a leadership over the companies that Argus ruled through its web of minority interests and its boardroom influence.

When McDougald died, the shares of Ravelston were split five ways, McDougald's widow, Jim, held 23.6 percent, and her sister Doris, Eric Phillips' widow, who lived next-door to her in Palm Beach, also held 23.6 percent through the Phillips' estate. Meighen held 26.5 percent, the largest single block, through two companies, Canadian General Investments and Third Canadian General Investments. Matthews held a small interest of just 3.9 percent. The heirs of George Black, his sons Montagu (Monty) and Conrad, held 22.4 percent through a company called Western Dominion Investment. Under these circumstances, with the widows of McDougald and Phillips out of the picture and uninterested in becoming involved, Meighen and Matthews — together with Alex Barron who was president of Canadian General Investments — assumed that the succession had passed to them. For a brief time, it had. But the youngest director of Argus, Conrad Black, aged thirty-three, had other ideas.

Black might well have waited for his chance. But he chose not to. One day after the death of McDougald, he had met Barron for dinner and told him that he would like to take up an executive post with Argus and have his brother, Monty, elected to the board. In the past, he and Barron had formed an alliance together, trying to persuade McDougald, an archconservative, that the time had come for Argus to play a more active role. Black was left with the impression that Barron would put his case to Meighen and the others. One week later, Black — never an early riser — arrived late at an Argus executive committee meeting to find that nominations for all the top posts at Argus had been made. Meighen was to become chairman, Matthews president and chief executive, and Barron executive vice-president.

The meeting had occurred precisely one week after McDougald's death, on March 22, and for Black it amounted to a declaration of war. Within a matter of weeks, he had devised a strategy to convert his own small stake in Ravelston into a winning hand and wrest control of Argus from the Meighen group. Crucial to the plan was the consent of the widows in Palm Beach; an alliance with them would provide Black with a 70 percent block of Ravelston, and an invincible position.

One important ally in his bid was an Argus director, Dixon Chant, who was an executor of the Phillips' estate. Another was Albert Thornbrough, who knew the widows well and was by this time disenchanted with the role that Meighen and Barron were playing at Massey. The widows were approached and won over and agreed to sign a voting agreement, an act which they later told everyone they regretted.

Within two days of the documents being signed, Black and Chant served Meighen with a compulsory purchase notice requiring him to sell out his 26.5 percent shareholding in Ravelston. Meighen was dumbfounded. Under the original shareholders' agreement, a provision had been drawn up that made it mandatory for a minority shareholder to sell out if the majority, for whatever reason, wanted him to do so. Black was invoking this clause. And he was doing so on the basis of his voting pact with Mrs. McDougald and Mrs. Phillips — both of whom did not seem to be aware that their alliance with Black would help to dislodge Meighen.*

Until now, little had surfaced publicly of the feud that was taking place. Ravelston was a private company, designed by McDougald to keep disputes between its partners away from the public gaze. Early in June, however, Meighen and Barron were forced to inform their shareholders at Canadian General Investments of what had transpired. Suddenly, the spotlight was turned on the disagreement at Ravelston, and its consequences for the future of Argus. Under the ultimatum, Black's Western Dominion Investment Company was compelling Meighen to sell all his Ravelston shares within six months.

The unwelcome publicity also brought the widows back into the fray. They were not happy with what was taking place in Toronto. Acting on advice, they had determined to do some-

---

* A reporter on the Montreal *Gazette* who phoned up Mrs. McDougald and Mrs. Phillips found them both surprised to learn what had happened. They explained that they had signed hundreds of papers since their husbands had died, presumably without inquiring as to the contents of all of them. "I don't think anyone is forced to sell anything," said Mrs. McDougald. In the same story, Conrad Black was quoted as saying the widows "knew exactly what they were doing."

thing about it. A Toronto real estate developer, John Prusac,* who had bought some land off McDougald a few years earlier, came on the scene as an adviser to Mrs. McDougald, and the two widows sought the counsel and help of General Matthews.

The plan they devised as a counter-strategy was simple and seemingly effective. Together, the widows would buy out Matthews' small 3.9 percent interest in Ravelston. In this way, they would gain a majority position of 51.1 percent and be able to undo their voting pact with Black. Personally they had no desire to intervene in the affairs of Argus, but they were disturbed about the controversy that had been stirred up and wanted to restore order and calm to the business empire their husbands had bequeathed them.

In Toronto, Black responded to the challenge at once. In April he had been made a director of Massey. On June 27, he was attending his third meeting of the board. During a break in the proceedings, Black went over to Massey's chairman, General Matthews, and served him with the same notice that he had handed to Meighen — a compulsory order to sell his Ravelston shares to Western Dominion Investment Company.

The document that Matthews was required to sign was different to that signed by Meighen in one respect. It did not contain the signatures of the two widows, who would not have given their approval. Black and his lawyer, Igor Kaplan, still had the voting agreement with the widows' signatures, and it was a ten-year agreement; in their opinion, there was no need for them to sign again. The serving of the notice on Matthews was not by itself decisive. But it did stop him from proceeding with his plan to dispose of his interest to Mrs. McDougald and Mrs. Phillips. Black meanwhile had set out to convince other Argus directors that the intervention of the widows would be disruptive; it would stand in the way of the company and prevent it from functioning effectively. On the board, he had two close allies and friends, both of whom had been business associates of McDougald, Nelson Davis of the N.M. Davis Corporation and Hal Jackman of Empire Life Insurance.

Ten days later, at the request of Black and Davis, a special meeting of the Argus board was convened. It was to be the final confrontation, and Black made careful preparations in advance. He canvassed the dozen or so directors who were not officially committed to either side in the struggle and at a caucus meeting beforehand assured himself that he had enough votes to replace all the Argus officers. As it turned out, these manoeuvrings

* Prusac's only previous involvement had been to purchase a piece of land off McDougald. Of his appearance on the scene in Palm Beach, Nelson Davis, a Black ally, remarked: "You can't suddenly decide to do business with anybody who walks down the street." The remark was made to the *Globe and Mail's* Hugh Anderson.

proved unnecessary. A deal was made with Matthews. In return 217 for allying himself with the Blacks, it was agreed that he would keep his stock.

The fight was over. Without Matthews' shares, there was no hope that the widows could strike back. A few days later, it was announced that the shareholding of the McDougald and Phillips estates would be sold to the Blacks for $18.4 million. Conrad became president of Argus. Monty became president of Ravelston.

During the summer months and into the fall, Black worked to consolidate his position and to co-opt into Ravelston a group of Canadian establishment figures who could provide both managerial strength and the large-scale financial resources he and his brother lacked; Nelson Davis, Hal Jackman, who already had a 9 percent interest in Argus, and Fredrik Eaton, chairman and chief executive of the Eaton department store chain. Their backing was to be crucial, for Black needed to do more than merely triumph in an internal power struggle. He also needed to impose his will over the diverse holdings of Argus by taking initiatives and moving in new directions.

A single-minded and tenacious exercise.

By gaining control over Ravelston, Black had pyramided himself into a position of power over four out of five of the Argus companies, Massey-Ferguson, Dominion Stores, Hollinger Mines, and Standard Broadcasting. The fifth company, Domtar of Montreal, was less likely to surrender. Despite Black's ascendancy over Argus, Domtar was squarely in the hands of his opponents, Meighen and Barron.* Nominally, Black had put himself in charge of almost the same block of Canadian business that McDougald had commanded, but his control was far more likely to be challenged, and far less solid financially.

To buy out the Meighen, McDougald, and Phillips interests, Black called on his wealthy partners to put up half the $30 million that was required. He and his brother had $7 million. The rest they raised through bank loans using a newspaper chain, Sterling Newspapers, as collateral. In return, the Blacks had to bring their partners in the newspaper venture into the family holding company, Western Dominion. The result of these dealings was to leave the Blacks with 80 percent of Western Dominion (their

* Black soon afterwards disposed of the 16.9 percent Argus interest in Domtar, selling it to MacMillan Bloedel of Vancouver for $67.5 million. In an interview with Dean Walker of Toronto's *Executive* magazine, he later declared that he had exercised no more influence at Domtar than "a little old lady with 100 shares."

newspaper partners, David Radler and Peter White, had 20 percent) which in turn had a bare majority of 50.5 percent control of Ravelston (the rest going to Black's wealthy friends).

Control over Ravelston did not, however, guarantee holding and keeping supreme authority over Argus. It had for McDougald, with his network of establishment friends and connections. But it might not for the Blacks. The legacy they had inherited included a block of Argus stock that Ravelston did not own. It was in the hands of Paul Desmarais of Power Corporation and consisted of 26 percent of the common shares and 59 percent of the Class C preferred shares acquired during Desmarais' celebrated takeover bid three years earlier. Desmarais was willing to sell. But it would cost the Blacks $80 million of which $65 million would have to be in cash, and the rest in a promissory note.

To raise the money, the Blacks resorted to an ingenious and complicated manoeuvre. They persuaded two banks, the Canadian Imperial Bank of Commerce and the Toronto-Dominion Bank, to take preferred stock in a newly launched Ravelston subsidiary and to provide them with the $65 million they needed in return. The dividend yield from this purchase for the banks was lower than interest on a bank loan, but it was made attractive to them because the dividends were nontaxable. In the case of Ravelston, the Blacks wanted to avoid a bank loan. The company's only income came from nontaxable dividends; it could therefore not offset any interest deductions against taxes, and could not afford the interest charges on the loan. As for the dividend that would be paid to the banks, it would be covered by the earnings Ravelston would receive from the Argus stock purchased from Desmarais.

By the end of the summer, Conrad Black had completed his power play and had manoeuvred himself into a position of control. It had been a single-minded and tenacious exercise, accompanied by a great deal of publicity which he had encouraged. Argus had come out from the shadows and was once more in the public eye as it had not been since the heyday of E.P. Taylor in the 1950s, and Black was being seen as a vital new force (the "boy wonder" of Canadian business, as he was to be dubbed by *Fortune* magazine six months later). With his victories, Black had assumed enormous power and an awesome amount of responsibility. And nowhere was this responsibility more evident than at Massey-Ferguson. On August 16, Conrad Black had been elected chairman of the board.

In the winter of 1977, several months before McDougald's death, Conrad Black had had luncheon with Victor Rice at the Maison Basque restaurant in Toronto. The two had never met before. The luncheon had been suggested by a mutual friend at Massey. Black, a director of Argus, was already getting himself into the newspapers, and he had been the subject of a magazine article that had speculated on his future as the heir to McDougald. Rice had read it. And the friend had suggested he should meet Black, telling him that they would probably get on well together.

The luncheon at Maison Basque lasted for several hours, and the two did get on well; both men were about the same age and, though they came from entirely different backgrounds, they had much in common. Black at the time had an office in Dominion Securities from which he operated his business interests. He was a director of Eaton's and of the Canadian Imperial Bank of Commerce as well as Argus. Rice had just been made the corporate vice-president for staff operations at Massey. Essentially, both of them were marking time, waiting in the wings for the opportunities that would soon present themselves. And since both were ambitious, they were frustrated with their roles: Black had tried and failed to get Argus on the move, Rice was battling to get some sense of reality into the management of Massey, and also failing. During lunch, the talk turned to Massey. But Black was not involved in the company at the time. For the most part, the lunch conversation ranged over the Toronto business scene with Black, forthright and entertaining, giving his views on the personalities surrounding Argus.

The luncheon made the two social acquaintances, if not friends. It also formed the basis for an alliance between them. Black was not in any way committed to Rice, or to pushing his career forward, but he had met him, knew him personally, and knew that he could work with him if the occasion presented itself.

Conrad Black had been born into the Canadian business establishment and had gone to the best schools. He had graduated from Carleton University in Ottawa and then moved on to law school at Osgoode Hall in Toronto. When he was told he had to repeat the first year after doing poorly, he decided to leave for Quebec and team up with a friend in the newspaper business, Peter White. His academic career continued at Université de Laval where he took a degree in Quebec civil law. Intelligent and well-read — an attribute that Black dates from his reclusive childhood when he read books rather than playing sports — he also became involved in politics. After moving to Montreal and enrolling at McGill University, he produced a weighty biography of Maurice Duplessis, premier of Quebec from 1936 to 1959. When a number of newspapers reviewed the book unfavorably, Black launched his own personal campaign against his critics. The newspapers were made to hear about Black's displeasure. Later, as his business career developed, the press treated him with great respect, taking an almost sycophantic attitude to everything he said and did; in one popular newspaper he was described as the "prince of tycoons."

By his early thirties, Black had established himself as an intellectual and something of a maverick. It was unusual for a man of his background to leave Toronto and immerse himself in Quebec politics and culture. Moreover, his biography on Duplessis was a polemical work that questioned the conventional view, popular in both Ottawa and Quebec City, that the Quebec leader had been a crass reactionary, standing in the way of progress and the natural aspirations of his people.

Black's unconventionality spilled over into his early business dealings. He was a rich man's son and did not have to strive to make money or to establish himself. Nonetheless, the task of making money, of founding a business and running it, had appealed to him as a challenge; much like any other university student, he had banded together with a group of friends to test his ideas.

Peter White was an assistant to Quebec premier Daniel Johnson and owned a small weekly newspaper. David Radler was the son of a Montreal restaurant owner. He had studied business at McGill and Queen's universities, then set himself up as a consultant to small businesses. The two had met at a by-election victory party given by the Union Nationale party in 1968. Black, who was studying law, also met White through politics. Together, Black, Radler, and White formed a partnership. Their idea was to exploit the one strength they had, White's knowledge of the newspaper trade, and see it they could build a successful business. They were buoyed in their enthusiasm for newspapers by the fact that many great Canadian fortunes, such as the Thomson's and Southam's, had sprung from small-town newspaper operations.

In 1969, the *Sherbrooke Record* owned by Toronto publisher John Bassett came onto the market. The asking price was a mere $20,000 for what was a none-too-profitable English-language daily in the heart of Quebec. Black, Radler, and White each put up $6,000 and bought the paper, then installed themselves as a management team with Black handling editorial, Radler advertising, and White production. Like any other small business, there was only one method that would work to generate the profits that they were looking for, and that was to cut costs. Within three months, the new owners had slashed the staff from forty-eight to twenty-four. The cuts were made in all departments, but they fell most heavily on the editorial side. The young entrepreneurs, it was said, became the initiators of something new to the newspaper industry, the three-man newsroom.

The gibes about the cheapness of the newspaper did not offend Black and his partners; they were looking at the bottom line, and the *Sherbrooke Record* was quickly becoming very profitable. Profits were then rolled into other newspapers, all of them small and local. The new acquisitions were financed by bank borrowings, with the loans often guaranteed personally by Black. Still, the company, Sterling Newspapers, would pay off the borrowings and promissory notes from the profits it was generating; Black was not using any of his inherited money to keep the company going.

In 1971, Sterling Newspapers switched its operations from east to west. The decision was made to sell out in Quebec and the Atlantic provinces and move to British Columbia. Six small papers were sold for $700,000, while the *Sherbrooke Record* fetched $1 million. Having bought the paper for $20,000, the trio had made $1 million on their investment in three years and then managed to sell it for another $1 million.

When it came to investing the money in B.C., Black bought three papers using the simple device of phoning every small-town newspaper publisher in the province and finding out which ones were prepared to make a deal. More acquisitions followed. By the time Black was ready to make his move for control of Argus, Sterling Newspapers was worth about $14 million and three partners had become four, with the addition of Black's older brother, Monty.*

The aspirations of Conrad Black, and his determination to succeed on his own, had their counterpart in the personality of Victor Rice. Just three years older than Black, Rice had not come from a privileged background. Nor did he have any academic qualifications. But he was intensely ambitious and he had resolutely set himself the goal of climbing upwards in a large corporation.

Rice had grown up in the suburban town of Ilford outside London, U.K., and had gone to a local grammar school. He was not too interested in studying, and at sixteen left school with the idea that getting a job at an early age would give him working experience and put him ahead of less pragmatic competition. The Youth Employment Bureau sent him to the

---

* Monty Black had come in following a deal in which Sterling acquired Dominion Malting Ltd., of Winnipeg in 1976. The company, once owned by E.P. Taylor, produces half a million pounds of malt a day, a third of total Canadian production. Though it had nothing to do with newspapers Monty wanted to buy it and subsequently he backed into a one-fourth interest in Sterling.

London Rubber Company, and they promptly hired him. A jubilant Victor returned home to tell his father that he had his first job as an office boy. His father congratulated him and then asked him what the London Rubber Company made. Victor did not know and returned to the company to find out. When he discovered they were in the business of making contraceptives, he decided to forego the job. Instead he found another office job at the huge Ford Motor Company plant at Dagenham.*

The switch proved to be a fortunate one. The U.S. company was in the middle of taking over its British subsidiary and revolutionizing business methods, bringing in new ideas and financial management concepts that had not been introduced in Britain before. Rice was a junior clerk but he managed to absorb the new techniques and get himself promoted quickly into accounting and financial analysis. He later moved to Cummins Engines, and then was hired by Chrysler. The British Chrysler company set out to recruit a number of bright young men who had graduated through Ford, and Rice was one of them.

In 1970, at the age of twenty-nine, he joined the Perkins Engines Group, Massey-Ferguson's subsidiary based in Peterborough, and was appointed the company's comptroller for North European operations.† Thereafter, he rose swiftly in the ranks as director of Finance, then director of Sales and Market Development. Within three years, he had emerged at the top, becoming deputy managing director of Operations.

From this post, Rice effectively ran Perkins. His boss, Monty Prichard, was often out of the office, travelling, or on sales trips, or putting together deals. In Peterborough, the head office in Toronto, Canada, was a remote and distant place, and Prichard had taken pains to keep things that way to maintain Perkins' independence. There was the occasional visit from Thornbrough. The Perkins executives would worry more about the arrival of Thornbrough's Rolls-Royce which would be parked outside the factory for all to see than about the visit of the chief himself. Labor relations at the British plant were as difficult as they were anywhere else in the country, and the sight of the boss in a Rolls-Royce was like a red rag to a bull to Perkins' shop stewards.

In 1975, much to his disgust, Rice was moved to Toronto

* Rice's first job as a mail-room boy paid roughly $700 a year. Twenty years later, as chairman and president of Massey, he was earning $390,000 a year.

† Rice was never reluctant to talk about his ambitions or persuade others of his ability. A colleague recalls attending a reception at Peterborough only six months after Rice had joined and being told by Sir Monty Prichard: "That man will be president of the company one day."

as comptroller of Massey. He did not want to leave Peterborough
or Perkins. And he did not have much taste for the political
job of comptroller; the appointment would set him against
the executive vice-presidents, and put him in a position of
having to stand out against some of the ambitious and free-
spending plans that the company kept pushing forward.

For three years, as comptroller and then vice-president of
staff operations, Rice was engaged in a head-on confrontation.
In any argument he was invariably the loser. Thornbrough
preferred to listen to his men in the field. Rice kept on pressing
the point, stating and restating his concerns about the financial
state of the company and its ability to weather a downturn in
its main markets. But he made little headway. The executive
vice-presidents were ranged against him, and the promotion of
any one of them to the president's chair would have led to the
end of his career with Massey.

Thornbrough gave him little support though he did acknowledge, at Rice's insistence, that the construction machinery
business was turning disastrously unprofitable and should be
disposed of. It was a small victory, but it was a start. And by
the spring of 1978, with Massey having to announce a $38 million
loss in the first quarter, Rice's message was beginning to get
through.

The troubles that Massey faced, and the emergence of Conrad
Black at the helm of Argus, were to bring Rice and Black together
later in the summer of 1978, and they were to establish a real
and effective working partnership. Together, they committed
themselves to turning the company around and pulling it back
from the brink. For all their common interests and shared
ambition, however, they approached the task from distinctly
different viewpoints. Black, the entrepreneur, the young man
following in the footsteps of Bud McDougald, saw Massey, as
McDougald would have, as an investment and an important
one that must be made worthwhile by being made profitable.
Rice, the company man who understood more about big
business, and the systems and structures that made a global
corporation function, saw that changes had to be implemented
and some radical surgery performed at Massey. Even so, it would
be no simple task to improve Massey's performance. The
company had wasted too many years going in the wrong
direction.

In the summer of 1978, these differences were not very
apparent. Both men realized the urgency of the task that lay
before them. Later, however, they were to create real problems
and result in a complete divergence of views. Ultimately,
the similarities between Conrad Black and Victor Rice were
to become less important than their differences.

While the contest for power preoccupied the directors and principals of Argus, the situation within Massey was going from bad to worse. Farm equipment companies have traditionally had to ride out rough cycles, struggling to meet demand when farmers are prosperous and eager to buy new machinery and endeavoring to get by when farm incomes go into decline.

Two years earlier, Massey had been enjoying boom times. But there had been ominous signs of a downturn in 1977 and, by the summer of 1978, it was evident that farm markets were deteriorating fast even as interest rates pushed up the cost of servicing the company's huge debts. Important markets in Brazil and Argentina had gone into a precipitous slump; U.S. and Canadian farmers were cutting back as farm prices and incomes fell; and poor weather conditions had blighted harvests in Europe the previous year, leaving farmers unwilling and unable to invest in new tractors and harvesters. These setbacks in the marketplace had created a big build-up of inventories. And they had led to a state of emergency within Massey as the company strove to cut its costs, reduce manpower, and trim production.

In June and July of 1978, Massey's board of directors had been unable to decide about anything. Their preoccupation was with events at Argus. Meetings of the board and executive committee were held in an atmosphere of confusion and animosity. Within the company, there was a mounting sense of alarm about prospects in the months ahead, and this was reflected in a board meeting in July when Rice was given the job of telling the directors how serious the condition of the company was. There had been intimations before that Massey was in real trouble. Thornbrough had announced that losses for the first six months of the financial year were running at $55 million. But the directors had been won over by a sense of false optimism. The company had been through tough times in the past when earnings had dipped for one or two quarters, and it had managed to stage a recovery.

The message Rice gave them came as a shock. Losses for the year would be substantial, he told them. Moreover, if the company was to survive and return to profitability, drastic action would be needed to slim down its operations and improve its balance sheet. To stage any kind of recovery, Massey would have to unload all its nonessential operations, take massive writeoffs in what was going to be a loss-making year, and concentrate on the profitable core of its business. Ultimately, the company would have to obtain new equity financing. Massey could not afford the interest expenses on its short- and

long-term debt. In 1976, these had amounted to $100 million.
In 1978, they would come in at $187 million.

The question of what action should be taken was intimately tied to another unresolved problem; who was to become the president of Massey and take charge of the rescue? Plainly, time was running short. If an outsider were to be found and appointed, he would need time to familiarize himself with the company and its problems. And it was doubtful that Massey could afford any kind of delay. Decisions had to be made quickly. The resignation of Peter Breyfogle had cleared the way for Rice, but, he was still regarded as young and inexperienced. In the organization and management committee, there was opposition to Rice. Far better, it was felt, to find a president who had experience running a $3 billion corporation and turning it around.

On August 16, Conrad Black was made chairman of Massey replacing General Matthews and he immediately moved to deal with the succession problem. Having won his battle for control of Argus, he had also managed to break the deadlock at Massey.*

"The company had to skim off the cream."

One of his first actions was to consult with the organization and management committee. Black wanted to appoint a president quickly, and he favored Rice. Thornbrough agreed. The committee, consisting of Thornbrough and a group of Massey veterans, senior corporate vice-president John Staiger, corporate vice-president for personnel and industrial relations Jack Belford, and corporate vice-president Wally Main, did not endorse the choice unanimously; Belford held out in opposition to Rice. But Black was able to let Rice know that the probability of him getting the job was high. On September 8, with the approval of the board — which still included Meighen and Matthews — Rice was made president and chief operating officer.

The personal rivalries that had plagued Massey for so long were not forgotten. Rice had come from the position of being a rank outsider. There were senior executives who felt that they could not work with him, and Rice had a similar antipathy to them. Just as the clashes within Argus had led to defections and resignations, so too, did the sudden elevation of Rice transform the senior ranks at Massey. Out of sixteen corporate

---

* Meighen and Barron resigned from the board before the end of 1978 to be replaced by Conrad Black's slate of directors which included David Radler, Dixon Chant, Fredrik Eaton, and Black's brother Monty.

managers at the top of the company in 1977, eight either resigned, or retired, or were forced out. In some cases, the partings were amicable. In other cases, executives who had been senior to Rice decided they could not remain and work under him. John Mitchell, a leading candidate for the presidency, resigned ten days before Rice's appointment was announced. Hugo Vajk, another executive vice-president and claimant for the presidency, resigned a few months later. Rice had wanted Vajk to stay on, but he decided against it. Jack Belford, who had opposed Rice, left the company within six weeks. Others such as John Staiger and Wally Main soon retired.

The departures were expected, and they cleared the way for Rice to promote his own staff. Having held the post of vice-president of staff operations, he had been able to judge the performance of middle-level executives. As he reorganized the company, centralizing its decision making in Toronto, key posts were given to executives who had been operating managers in the regions. From now on, Massey was to be far more tightly controlled from head office. The baronial system, encouraged by Thornbrough, was quickly abolished.

The appointment of Rice as president was a surprise to many within Massey. For a number of executives who had felt frustrated by the blindness of the previous management, it carried some hope that the company would finally begin to tackle its problems. But the task was a daunting one. On the same day Rice's appointment was announced, Massey reported a net third-quarter loss of $90 million, the worst performance in its history.

A statement put out in Thornbrough's name announced that "action programs, which required substantial management effort and incurred large one-time losses, are now identified and will return the company to profitability in 1979." But shareholders, and the investment community, were far from reassured. The company intended to find a buyer for its construction machinery business but had not done so yet. It was endeavoring to reduce inventories, cut its labor force by 9,000, pare down manufacturing costs, and extricate itself from unprofitable operations in Brazil and Argentina. Still, sales were flat or falling in all major markets, while in some markets, the slide during the third quarter had been disastrous. Sales in Europe were down 5 percent, in South America down 22 percent, and in Australia down 24 percent. In North America, the company continued to run into quality problems with its newly launched line of large tractors, and sales were showing virtually no real gain at all.

Rice's view of this situation was that Massey must continue to retrench. The company had to skim off the cream, he told

Black, and this meant taking tough decisions; firing a lot of people, trying to sell off businesses that had been hastily acquired, closing down factories, and cutting costs. Providing markets for farm equipment recovered to normal levels, this process would begin to achieve results and a return to profitability. The second part of the strategy would be to undertake a reconstruction of the company, deciding what sectors of its business would be more important in future and developing these strengths. Once Massey began to turn around, and it had been demonstrated that the company was going to be profitable again, steps would have to be taken to secure new equity. Inevitably, any equity financing would involve Argus. Black as chairman and principal shareholder would have to make the decision on the terms and timing of a public issue, as Argus had traditionally done in the past.

The priority, however, was to reorganize and slim down Massey, to take as much of the losses as it was possible to take in 1978, and hope that earnings would improve thereafter. Rice and Black were both agreed on this course. And so was Albert Thornbrough, who remained on as deputy chairman and chief executive officer. Thornbrough saw his role as an elder statesman, advising and counselling his young successors, and he continued to occupy the spacious presidential office.

In reality, the new chairman and the new president were heeding their own advice and taking the company in a diametrically different direction, shrinking instead of expanding, destroying and then rebuilding the management structure, taking Massey away from the peripheral lines of business that had been thought to represent the wave of the future. Moreover, Rice and Black were moving as quickly as possible in order to prevent a worse calamity from befalling the company that Thornbrough had led for so long.

14
"We Can Do It"

*The single objective that Rice had set himself
was to make Massey profitable again. He was
determined that the company should come
through since, only by demonstrating its
capacity to make money, could he move onto
the next stage; the reconstruction of the
company and the injection of new equity to
put it on a sound financial footing.*

Within Massey, there was a great deal of concern about
the barrage of publicity that surrounded the company from the
latter half of 1978 and into 1979. The figures that told the story of
Massey's plight were a visible reminder of how seriously things
had gone wrong. After making provision for reorganization
expenses and factory close-downs, the company lost $257 million
in a single year, the equivalent of a $14.53 loss for every common
share. It was also forced to renegotiate debt obligations that
could not be met. Certainly, this slate of bad news could not be
swept away or hidden from the public. Never before in corporate
history had any Canadian company done as badly. But
demoralized as they were by this state of affairs, many Massey
executives reacted angrily to the treatment they were being
given by the newspapers and the media around the world. And
long-serving company men, Thornbrough among them, felt
that a lot of the negative publicity that Massey was being given
came out of the high public profile accorded Conrad Black and,
to a lesser extent, Victor Rice.

Black was extremely news oriented. Articulate, outspoken,
and given to philosophizing about Massey's problems, he
attracted newsmen and news coverage and seemed to revel in
the prestige and prominence his new position gave him. He was
elevated to the title of Canada's "Businessman of the Year" by
the *Globe and Mail*, though the paper noted that the attention
Black was receiving was "as a maker of news, not a maker of
companies." When he moved his office to Massey, Black was

asked by the public relations men at the company whether they could assist him in handling relations and interviews with the press. Black replied in the negative; he was himself a newspaper publisher and adept at such things.

Often, the statements that were made seemed to have a positive impact. They were intended to underline Black's commitment and the energy and drive that Rice was putting into reorganizing the company. But, inevitably Massey was being portrayed as a company with almost insurmountable difficulties — the problem child in the Argus litter. Sometimes the statements were quite equivocal, as when Black, in an interview with *Executive* magazine, spoke of "shaky-kneed and sweaty-palmed" bankers coming to find out the status of their loans, or declared that his impression of Massey was "not that there were knavish people around but there were certain exaggerations of optimism."

Black was certainly being candid. But the effect of such statements on a company of the global size and diversity of Massey was not easy to contain. Massey was in bad trouble, and the people who were running the company were saying as much. Not unnaturally, as the difficulties mounted, so did the volume of unfavorable publicity. To the people who had always attempted to shield the company and protect its good name, the outpouring of bad news became in itself a cause of concern.

Behind this worry was a legitimate fear. The strength of any farm machinery company rests on its network of distributors and dealers around the world. More than in most industries, the ability to make a sale depends on the trust that has been built up between seller and buyer. Massey had, in any case, been neglecting its dealers; as business had dropped off, they had been running down inventories to obtain additional cash, but had still found themselves squeezed with very little support from head office. Now dealers were having to contend with an outburst of publicity about the precarious state that Massey was in.

John Ruth, who was to be appointed the company's vice-president of marketing, recalls visiting friends who were Massey dealers in Europe. He found many of them losing money, and complaining about the treatment they were getting; the company was often making decisions that were not in their best interests, withholding money for warranties or delaying payment. Brian Long, a young manager transferred from the three A's regional office to take charge in France, found product development had lagged. Massey had been well-established in France, outselling its chief rival Renault, but in the space of a few years its share of the market had shrunk from 22 percent to 13 percent.

In the largest market of all, the U.S., dealers were openly voicing their discontent. A new line of tractors, in the over 100-horsepower category, had been plagued with transmission system breakdowns.* Buyers were having to take them back for repairs time and again. They were complaining to the dealers and telling other farmers about the unreliability of Massey's large tractors. The engineering and manufacturing staff were committed to correcting the defaults and repairing the tractors, but it was an expensive undertaking. Needless to say, the situation did nothing to improve Massey's reputation in a market where it was struggling to stay on terms with John Deere and International Harvester.

To.these complaints and grievances was added the unfavorable news about Massey's financial state. Any farmer going into a John Deere dealership in Iowa or Illinois would find, clipped to the notice board in the showroom, the latest article from the *Wall Street Journal* or the *Des Moines Register* or *Chicago Tribune* about Massey's problems and its failing financial condition. The message was clear: A farmer who bought an expensive piece of farm equipment from the neighborhood Massey dealership did so at his peril.

The same story was repeated wherever Massey did business. The publicity could not be ignored. Nor could it be countered, as the bad news and the red ink continued to flow. Within the industry, Massey was identifiable as the company with the big problems, the company which might not survive and was therefore risky to do business with. Any report about the company in the press was prefaced with descriptive adjectives: Massey was always described as "financially troubled" or "ailing" or "loss-making."

For the dealers, who were disgruntled anyway, the job of selling a piece of sophisticated farm equipment that might cost $40,000 or $60,000 was made more difficult. And this prompted many of them to pursue other options. Massey might not lose a dealership, but it would find that one of its dealers was hedging his bets and deciding to sell Allis-Chalmers, or Case, or some other manufacturer's equipment as well.

At head office in Toronto, this trend was of great concern. And it almost certainly precipitated Massey's second big mistake. Aware that Massey's reputation was on the line, Victor Rice became increasingly strident about the company's ability to turn itself around and make a profit. At a luncheon meeting with the press — who were regaled with an expensive menu of steak and Domaine de la Tour wine — Rice predicted that Massey would be profitable by the following year.

The prediction was made with an eye to reassuring all

---

* Massey had to send out over 100 service engineers who spent months on the road repairing tractors and replacing faulty 24-speed transmissions. It also had to train service people in all its North American dealerships to handle the special problems associated with the transmission and steering columns.

the people Massey did business with, and its unhappy shareholders who had seen their shares plummet from $32 in 1976
to $12 in 1978. Since Rice intended to slash away at the
company's unprofitable operations and reduce costs drastically,
the forecast was not unreasonable. Still, this statement, and
others, created the expectation that Massey would do substantially better, an expectation the company manifestly failed to
live up to. To succeed, Massey needed a worldwide sales
recovery. And this never materialized.

The question of credibility, and of fostering the right public
image, has traditionally been a difficult one for corporations
in trouble. Inevitably, companies react to pessimistic news by
being overoptimistic. At Massey, the problem was compounded
by the amount of personal publicity that was swirling around
Conrad Black, and the natural enthusiasm that Victor Rice had
for the task that lay ahead. The company was alternately a
source of pessimism and optimism, and it was consistently
making news on both counts, often to its own detriment.

The impetus for change was coming from the top.

The corporate headquarters of Massey-Ferguson are now
situated in a modern glass-fronted building in downtown
Toronto. But for twenty years the company was located in a
drab building, above the local offices of the Sun Life Assurance
Company. The head office had moved there in 1961 from the
King Street factory and, in the Thornbrough years, the
ordinariness of the setting was both intentional and appropriate.
Massey did not stress its Canadian roots. It saw itself as a global
corporation, functioning in nearly every country in the world
and requiring only a modest head office staff. And though
Massey regularly appeared in the lists of the top ten corporations
in the country, it was the least known and least publicized of
Canadian companies.*

On the top floor, the executive suite of offices consisted of
a boardroom, a smaller office and two large offices, one facing
north towards Queen's Park and the Ontario parliament buildings
and occupied by the president, and the other the chairman's
office, looking south and east towards the downtown business
district and Lake Ontario.

---

* In Canada, Massey ranks seventh in assets (behind Canadian Pacific, Bell Canada,
 Alcan Canada, Imperial Oil, Inco, and Noranda Mines), and eighth in sales (behind
 General Motors, Canadian Pacific, Ford Motor Company, Imperial Oil, George
 Weston, Bell Canada, and Alcan Canada).

When the move had been made, Massey had been governed by the strong personality of Colonel Eric Phillips. The colonel had been chairman and chief executive officer, and he had watched over the Argus interest in Massey with great care. In the early days, when Thornbrough had just been appointed president, there was one occasion on which he had to phone the colonel for permission to buy a desk for the London office. Phillips had moved with the company and had installed himself in the chairman's office. But the modernity of his surroundings had depressed him, and he had the whole office redecorated and fitted out with panelling to resemble the opulent premises of the Argus head office on Toronto Street.

After Phillips died in 1964, the chairman's office was left vacant. For a brief few weeks before his death, McDougald had taken on the duties of chairman, but he never occupied the office. Instead the whole top floor was given over to the president, Albert Thornbrough. The rich panelled office was occasionally used for meetings, and Thornbrough would pose there for pictures for the annual report. Otherwise it remained empty.

When Conrad Black took over, the vacant chairman's office became a symbol of what had gone wrong at Massey. For years, there had been nobody on hand to represent the interests of the shareholders or to exercise a veto over the ambitious spending plans of Massey's executives. Since the death of Phillips, Argus had neglected its responsibilities.

Black was determined to correct this. And as he made public statements about almost everything, his elevation to the position of working chairman was given prominence. Shareholders were told at the annual meeting in March, 1979, that there had been a change in the style and substance of the company; Massey now had a functioning chairman, something it had not had since 1964. Black would occupy Phillips' office. Four days earlier, as a sign of his intention to run things, he had taken the title of chief executive office from Albert Thornbrough.

In the space of six months since the overthrow of the Argus old guard, there had, indeed, been a lot of changes at Massey. The exodus of senior executives had gone on, and they had been replaced by Rice's nominees, or were not replaced at all. Steps were being taken to choose new vice-presidents who would take responsibility for strategic and product planning, marketing and manufacturing. But the main thrust had been relentless and ruthless cost-cutting.

The strategy was to discard unprofitable or "noncore" operations and return Massey to its profitable "core" business base, consisting of the six major combine, tractor and diesel

engine factories in Canada, the U.S., Britain, and France.*
Among the marginal businesses were those that only a few years
earlier had seemed promising acquisitions; the diesel plant in
Canton, Ohio, was closed, while losses at the West German
construction subsidiary, Hanomag, were trimmed and efforts
made to sell it off.† Massey would also dispose of money-
losing plants at Akron, Ohio, and Kilmarnock, Scotland, and
sharply curtail its operations in the Third World; pulling back
in Iran; selling interests in Mexico, Argentina, and Brazil; disposing of its stake in Spain's largest tractor company, Motor
Iberica; and taking a minority instead of majority position in
its South African subsidiary.

Targets were set for trimming the work force by 12,000.
And the desire to cut costs and manpower reached into the
head office. Rice's new way of doing things gave more power to
executives in Toronto. But head office had to show it could
operate more efficiently with fewer people. Even senior executives
were liable to find their jobs amalgamated or abolished. The
decisions were tough ones to make, and they were not without
their cost. Could staff loyalty be counted on? Was it wise to pull
back in Latin America, a potentially rich market for the future.
Should Massey relinquish its major share in a South African
company that had always been a source of good profits? The
ultimate argument in favor of being ruthless was that the
company could not afford half-measures. If it was to hold its
creditors at bay, and find the money to make its interest payments, sacrifices had to be made. Rice calculated that the cuts
would generate $150 million in cash, enough to keep Massey
solvent for the time being.

Behind the scenes, other measures were contemplated. The
Perkins engines division in Peterborough was regarded as the
best hope Massey had for the future since the energy crisis had
created fresh opportunities for Perkins, and its products.
Secretly, Massey officials in London opened talks with the Labor
government of James Callaghan about selling a minority interest
in Perkins to the government-controlled National Enterprise
Board (NEB). The talks ended in failure.†† The NEB was interested

---

* These were to be located in Peterborough and Coventry, U.K.; Detroit in the U.S.;
Toronto and Brantford, Canada; and Marquette/Beauvais, France.

† Massey did not finally end its ill-fated involvement in the heavy construction
machinery business until it sold Hanomag to IBH Holding AG of West Germany in
November, 1979. From the date of its purchase in 1974 (for $45 million) to its eventual
sale, Hanomag ranked as the company's worst error; in 1978, a writedown of
$43.4 million was taken to cover the expense of reorganization.

†† At various times, both Conrad Black and Victor Rice went to London to talk with
politicians and top British civil servants about selling a part of Perkins.

in the welfare of Perkins since the company had always made a substantial contribution to British exports. But the amount it was prepared to offer was ludicrously small. Black and Rice decided the offer was inadequate, and did not pursue the matter further.

In Toronto, the impetus for change was coming from the top. Black as chairman and Rice as president were in agreement on the broad terms of what had to be done. When it came to the operations of the company, Black did not interfere. Rice worked with one of his senior finance men, Helmut Mack, to draw up a blueprint for Massey's core business — the profitable heart of the company's operations — which must be left untouched. Tragically, on January 29, 1979, Mack was to die in an air crash. Newly appointed to be the director of finance for North American Operations, he was returning from Des Moines to his home in Toronto for the weekend. The hired Learjet in which he was travelling crashed in Detroit. Two Americans were killed, and so was the president of Massey-Ferguson Industries in Canada, Bill Murray. The loss of two senior experienced executives was one more blow for a company that had already suffered a great deal of adversity.

Amidst all these difficulties, Rice pursued a single objective — to make Massey profitable again. He was determined that the company should come through, since only by demonstrating its capacity to make money could he move on to the reconstruction of the company, and the injection of the new equity needed to put it on a sound financial footing. The struggle was going to be an uphill one, and to dramatize it Massey's executives were enrolled in a "We can do it" campaign, complete with buttons and stickers. Even as management savagely cut and trimmed away at its staff, morale had somehow kept high.

The key concept that Rice had to introduce was that executives would be accountable and be judged on their performance. While this might have seemed to be a normal requirement in most companies, the laissez-faire atmosphere of the Thornbrough years had created an altogether different style. If a vice-president or manager of an operating unit had been faced with a problem, he had looked around to find someone else to blame it on.

When, three months after becoming president, Rice learnt that North American vice-president Ken Glass had underestimated the losses that were projected for the U.S. and Canada, he travelled back from Europe to Des Moines and fired both Glass and his finance director on the spot. The mistake could have been a costly one; Glass's projection was for $20 million in losses rather than $54 million. In the past, a little juggling of the figures to lessen the bad news had been standard practice.

Rice stressed to his executives that he wanted to hear the worst.
Another executive to depart was Lee Elfes, a design engineer
who had run Massey's advanced engineering and design centre
in Detroit and had been held responsible for the problems
surrounding the large tractor line. Like other Massey executives
in the past, Elfes had run a one-man show and had pushed
his own ideas. The company's belated entry into the large-
horsepower tractor market had been a dismal failure, and he
was made accountable.

Massey was kept going by Rice's nervous energy and remarkable sense of commitment.

The immediate effect of Rice's promotion and the decisions
that he took was to destroy the management structure that had
been in place. The group of managers below Thornbrough was
composed in equal parts of company veterans who decided to
retire, or executives who felt they could not work under Rice,
and either left of their own volition or were fired.

For a time, Rice stood in command of a company which
had a string of inexperienced vice-presidents, all of whom
reported directly to him. Since Rice managed to get by on very
little sleep, worked sixteen hours a day, kept himself going by
chain-smoking L and M cigarettes and drinking strong black
coffee, and would respond to every query and attempt to solve
every problem often by phoning across continents in the middle
of the night, Massey was kept going by his nervous energy
and remarkable sense of commitment.

Things could only operate like this for a time, however.
Gradually, the new hand-picked team of corporate managers
began to take shape and assume more responsibility. In
November of 1978, the first batch of senior appointments was
announced and, from then until the summer of 1979, the
company was being reorganized at the top and changed entirely
at Rice's behest.

One survivor from the past was Douglas Barker, the
treasurer, who had a background that included law degrees
from Cambridge and Harvard and who was a member of the bar
in England, Ontario, and Manitoba. He continued as treasurer,
and was made a corporate vice-president. Other middle and
senior managers who were at head office or were brought in
to run things included Michael Hoffman from Perkins; Darwin
Kettering, a midwesterner with roots in the corn belt, took
charge of North American Operations; Don Douglass, who had

been on John Mitchell's Americas staff, was moved to head World Export Operations; Vincent Laurenzo, a graduate of Notre Dame University and a former Ford executive, became comtroller; Jim Campbell, who had started off on the production line in the Kilmarnock plant, was given responsibilty for European Operations; Ralph Ramsey, a South African, who had been in charge of Massey's project in Poland, took over engineering; and Phil Moate, a longtime Massey executive, became director of Organization and Employee Relations and had to preside over the cutbacks that were taking place.

The chain of command as it existed before had been elaborate, with decisions filtering down from Thornbrough and broad powers being delegated to the regions. Now, there was virtually a single line of authority, with eleven vice-presidents reporting to Rice, and major decisions to be made every day.

Fitted into this scheme were several new positions. These had been created by Rice to overcome what he saw as the shortcomings of the Thornbrough era, and to stop the company from becoming decentralized and divided. John Ruth, a graduate of the U.S. Military Academy at West Point and John Deere's sales chief for Europe, Africa, and the Middle East, was recruited to be vice-president of Marketing with worldwide responsibilities, a post that had existed in the 1950s and 1960s only to be eliminated later. Another outsider, David Sadler, was brought from White Motor Corporation to take charge of manufacturing. As it turned out, Sadler did not stay with Massey long; he moved on to International Harvester eighteen months later. Both these appointments carried handsome salaries of $200,000. Brian Long, who had been manager in France and had a background as a mining engineer, a Harvard MBA, and who was also a fluent linguist was brought to Toronto to head up Strategic and Product Planning. Many of the failures of the past could be attributed to the lack of any co-ordinated plan. Massey had grown according to the whims and fancies of its senior executives.

Another newcomer was Michael Cochrane. He had helped wield the axe at two troubled Canadian companies, Air Canada and Reed Paper. Cochrane, who became vice-president of Planning and Business Development, did not last long. He was the first of the inner group of vice-presidents to leave when the troubles mounted in the spring of 1980. (He went to Northern Telecom. Others to leave subsequently were Sadler, and company secretary Derek Hayes, who went to Shell Canada).

For the most part, however, Rice kept his team intact. And he did it despite the pressures and strains that the company was going through. As it turned out, 1979 was not destined to be a good year. Markets were still sluggish. The British pound began to rise under the influence of North Sea oil, driving up costs in the company's chief manufacturing base. In the U.S., the company continued to wrestle with the quality problems of its large-tractor line, and its share of the market actually fell.

The quarterly earnings figures showed progress was being made, and Rice continued to make optimistic statements. Massey had turned the corner. In May of that year, he had become convinced that things were under control. Inventories were being held back; manpower cuts were making the company a great deal more productive; and Massey's overall performance was showing a steady improvement. Still, it was hard to convince others. In the first few months of the year, investors had caught some of Rice's enthusiasm and financial analysts had begun to recommend Massey's shares again. The share price rose from $10 to $15.50 in the expectation that the company would turn itself around quickly. When quarterly statements came out showing only modest gains, the share price dropped back again. It was becoming apparent that Massey might just break even in 1979, but it would not stage much of a recovery.

The big worry was the state of the balance sheet. At the end of the third quarter, the company had shareholders' equity of $603 million against long-term debt of $625 million, compared with figures one year earlier — when Black and Rice had taken over — of $652 million in shareholders' equity and $674 million in long-term debt. However, the shareholders' equity had been raised by an extraordinary $95 million tax gain in Britain, the result of a tax change which had allowed Massey to reverse deferred income taxes created by inventory tax relief in Britain.

On the short side of the balance sheet, it was apparent how much the company was depending on the good will of its bankers and suppliers. Bank borrowings had risen to $733 million from $560 million while accounts payable and accrued charges came in at $748 million against $585 million one year earlier. In all, current liabilities had climbed while current assets had stayed at much the same level as the previous year. And, while it was current liabilities had climbed while current assets had stayed at much the same level as the previous year. And, while it was true that inventories were being kept down, weak markets in North America had pushed up receivables and dealer stocks.

Given this performance, it was difficult to inspire confidence in the future, although Rice had not the slightest doubt that a breakthrough was being achieved. His natural buoyancy and ebullience, and the energy that he brought to the task, were focused on promoting Massey and on pleading its case publicly. Through most of 1979 and into 1980, a kind of ritual was played out. Massey would produce its quarterly earnings statement. Investors and stock market analysts would respond to the news unenthusiastically. Then, Rice would appear on the scene, anxious to do battle. The figures had been wrongly interpreted, he would claim. In reality, Massey had achieved a dramatic turnaround, perhaps one of the most spectacular in corporate history.

At the end of the financial year, Massey's figures included a change in accounting methods. The company had reverted to the industry's normal practice of reporting sales in North America at the *wholesale* level rather than as sales to the farmer. This switch further confused the issue, even as it boosted net income by more than $20 million. When the stock market reacted negatively, Rice stepped in once again to clear things up and claim a victory. In truth, Massey's income statement was so loaded down with surplus information — besides the new accounting method, there were provisions for reorganization expenses, losses from discontinued operations and foreign exchange translations — that investors had every reason to be bemused and a little doubtful.

During its turnaround year, Massey had scraped together an operating profit of $30 million as against an equivalent loss of $133 million in 1979, and it had done this on sales of close to $3 billion. Obviously it was an improvement. Equally obviously, the company had done nothing to alter its precarious state or to show that it could generate reasonable profits at any time. Leveraged to the hilt and making no money, it was plagued by a desperate shortage of cash. Money was needed to pay down short-term bank debts and reduce a crushing interest bill which, in 1979, had amounted to $200 million. And money was something that Rice and Massey-Ferguson did not have.

The plan, as it had been conceived by Black and Rice, was for the company to demonstrate that it could be made viable. After that, it would attempt to raise new equity and refinance itself. The questions that had gone unanswered were: when might the company be considered to have turned the corner? and when might an equity issue be contemplated? While Rice was in charge of executing the turnaround, it was the responsibility of Conrad Black and Argus to make the moves on the financial side.

In part, this was a traditional role. Argus had always functioned as the protector and guardian of Massey's interests when it came to financing. Despite its comparatively small stake in the company, Argus had never relinquished its grasp on Massey or the presumption that it should dictate to management. When Argus had acted, it had supposedly been acting on behalf of all Massey shareholders, although in reality the interests of Argus came first. And this was to be as true in the case of Conrad Black as it had been in the case of his predecessors, E.P. Taylor and Bud McDougald.

Within Massey, Black had caught the attention of the senior staff with his prompt choice of Rice to lead the company. He had then endorsed the approach that his new president was taking, backing him to the hilt even when hard choices had to be made. (Within days of Rice's appointment, decisions had to be made to pay off 1,500 workers in Britain at a cost of $30 million.)

Such economies were necessary. Black had been approached by U.S. bankers, led by Continental Illinois of Chicago and Citibank of New York, and warned that Massey was in a critical condition. As a member of the board of directors of the largest lender of all, the Canadian Imperial Bank of Commerce, Black had been advised by the bank's chairman, Russell Harrison, that Massey's debt situation was deteriorating. So he had plenty of awareness of the problem, and every reason to align himself behind Rice. His taste for self-dramatization had led him to make numerous statements about Massey, and this had reinforced his commitment to the rescue effort. His own reputation was involved in what was being done.

When Black moved his office to the Massey building, it was to take on the task that Eric Phillips had once set himself, that is, to be a working chairman. It was true that the offices of Argus were being redecorated at the time so the move was a convenient one. And it was also true that, for much of the day, Black dealt with the business of Argus.* Still, he would frequently take time out to discuss the progress of the company with Rice, and when the two of them were in Toronto, these meetings were held on a daily basis.

Black did, however, have a different view of Massey. He was engaged in building a power base at Argus, and in endeavoring to do this in a new way. The public face that his predecessors had presented to the world had been rather more impressive than their private grasp of power; like an old general, Bud McDougald had held things together in his last years on the basis of past glories. The reality was that Argus had a minority position in a number of companies and, while this had been sufficient for McDougald to maintain his power base, it was not going to be adequate for Black. Moreover, for as long as Argus

* One estimate by Victor Rice is that Black worked on Argus business about 85 percent of the time while his office was at Massey.

remained a passive investor, it was dependent on the dividend income it received from others. Not only was Black's position open to challenge, but it was also a vulnerable one.

His ideas of what could be accomplished were rooted in his own experience of small business.

The solution, as Black saw it, was to remake Argus and turn it into an operating company. The experience at Domtar, where he had been frozen out by Meighen and Barron, showed just how weak his position was. Black therefore moved to establish a firm control over another minority holding, Hollinger Mines, allying himself with other shareholder groups, and making this the centrepiece of a new-look Argus; an operating company with a debt free balance sheet, taxable royalties from its iron ore mining operations, and a great deal of available cash. In addition, Black was keen to increase his interest — or that of the renamed Hollinger Argus — in two profitable companies which were part of the Argus stable, Standard Broadcasting and Dominion Stores.

Black's strategy took shape during 1979. By the end of that year, Massey found itself a little further removed from Ravelston Corporation, the controller of Argus and holder of the 16.4 percent interest in Massey.

Interposed between Massey and Ravelston was Hollinger Argus. The branches on the tree that led back to Conrad Black had become a little more numerous, and a little thinner. Black's holding company controlled Ravelston, which had a 52 percent interest in Hollinger Argus, which had a 77 percent stake in Argus which, in turn, held the Massey shares. Moreover, compared with other Argus holdings, Massey was plainly and obviously less favored. The Argus interest in Standard Broadcasting had been raised to a majority position, and the interest in Dominion Stores had gone to 37 percent from 23 percent.

These important moves that Black was making were designed to make his position less open to challenge. In the case of Massey, he was prepared and willing to talk in general terms about increasing the Argus position, but he had never felt any obligation to do so. Nor did he perceive it as part of the grand strategy in which he was engaged. Providing Massey could show itself to be worth supporting — and that would depend on an improvement in its performance which was never specified — then, at some point in the future, Argus would make an additional commitment.

had abrogated to himself all power over deciding Massey's
fate. Any discussions about the financial future of the company
were held between Black and the other parties whether they
were bankers or underwrites. In these matters, Black acted as
the owner. Rice was regarded as the manager who could offer
advice and information, but had no authority to make decisions.
In the past, Argus had traditionally behaved this way, and Black,
as a student of history, felt the practice should continue.

From the standpoint of Massey and its management, the
relationship was obviously unsatisfactory. The company had to
improve its standing so that it would win Black's support and
be considered a worthwhile investment for the future. But what
exactly would please Black and draw an outright commitment
from him? To that question, nobody apart from Black had an
answer.

There were difficulties, too, in Black's wish to involve
himself in the company. He did not interfere with day-to-day
operations and communicated his wishes through the president
and chief operating officer whom he had appointed, thereby
fulfilling the role of an active chairman without being overbearing
or disruptive. Still, he had clear ideas about what should be
done, and these increasingly drew him into conflict with Rice.

The problem was one of perception, and of background.
While Black had scored a stunning coup in his takeover of
Argus, the victory had come against weak and dispirited
opponents. Prior to that, he and his friends had built up their
newspaper group by assembling together a few small businesses,
which they had then made profitable by consolidating into
even smaller units. Sterling Newspapers at its zenith had fewer
assets than a Massey subsidiary in Italy or Iran. Black's background was in small business. Although he was in command of
Argus and therefore a force to be reckoned with on the Canadian
business scene, he had never before encountered a global
company of the dimensions of Massey; nor did he have any
experience of the problems that size and diversity might create in
managing a multinational and in getting people to act once
decisions had been taken.

Black wanted to move quickly. He wanted to see results.
He was anxious that the turnaround should be accomplished
both because he had put his personal reputation on the line
and because he was impatient to reconstruct Argus. Massey
was the weak line in the chain. While it continued to suffer one
setback after another, there were limits to the way Black could
remould Argus. Partners who would not object to being
associated with Hollinger Argus and its profitable ventures
would balk at a business association that tied them to Massey.

And Argus could not disassociate itself from the company.

In the early days, there were no disagreements. The priorities were straightforward and they meshed with Black's ideas of moving quickly and effectively. As a result, in the initial period Black found it easy to approve of what Rice was doing. It was necessary to cut back and to make reductions if the company was to be turned around, and this was an aspect of business that Black had had personal experience of in the past, albeit on a smaller scale. In meetings with Rice, he would often cite his management experience at Sterling Newspapers. His overall view of Massey was that its business had become unnecessarily complicated. The task to be undertaken was to rationalize and simplify.

Obviously, there could be no argument on this point. Many of the troubles the company faced could be traced back to its willingness to get into peripheral lines of business, in which it was never competitive. As a result, it had neglected its main line of business. In a sense, management under Thornbrough had perpetuated the style of the 1960s, with its emphasis on diversification and growth, and had carried it into the 1970s — unaware that they were being overambitious and that the business climate was changing.

Still, Massey remained a huge global company. Even after the cuts, it had forty plants in eleven countries that it owned itself, another twenty plants in ten countries in which it was engaged in joint manufacturing ventures, and eight other subsidiary companies and licensee arrangements in a further fifteen countries. It had sales of $3 billion, and 56,000 employees around the world. To bring about change in such a large organization, and to implement decisions taken in the boardroom or in the president's office, was a slow and time-consuming process, particularly when it involved demolishing one structure of management and replacing it with another.

Black seemed to have little understanding of this. He felt that Rice was dragging his heels and failing to act decisively. He also had very clear ideas of the need to trim back executive staff and prune executive salaries, and this caused further differences. The company, in Rice's view, could not afford to lose its more experienced executives, and it had to pay them properly; for a company as large as Massey, salaries of $200,000 were not excessive. Moreover, since the company's troubles had been so well-publicized, corporate headhunting firms were often phoning up senior executives and trying to persuade them to quit Massey and move to another company. Staff had to be paid properly to be retained.

The two of them had started out with a great similarity of views. They also had compatible personalities and continued to

be able to work together; both were young, iconoclastic,
anxious to get on with the job, and impatient for change. Nonetheless, between the summer of 1978 and the end of 1979,
differences of opinion began to surface more frequently. With
his other interests that took precedence over Massey, Black
would not be drawn into an actual commitment on financing
and on strengthening Argus's obligation to Massey. Black's
ideas of what could be accomplished were rooted in his own
experience in small business, and in his wish to see a quick
solution to Massey's problems. Plainly, there were limits to his
staying power when it came to involving himself in the company.

On August 23, 1979, trading in Massey shares was halted
in Toronto and New York, while a statement was put out by the
company on its plans for a three-part equity issue. The proposals
as they had been drafted by Black included a "substantial placement" of convertible preferred shares with Argus, a public issue
of participating nonconvertible preferred shares with warrants,
and a private placement of nonconvertible preferred shares. The
terms and the amount of each of the issues were still to be
finalized, but the sum involved would be between $300 million
and $500 million.

The plan served to clarify Black's intentions even while it
did nothing to clear up the question of when Massey would
receive the additional equity. The announcement was timed to
coincide with third-quarter earnings figures and, since these
showed no great improvement, it was plain the company would
have to wait before tapping the market. Black had devised a
refinancing scheme which was in line with his thinking that
Argus must have more voting control over the companies it
invested in, and designed to reward him if things went well. By
taking the preferred shares, convertible at $15 into common
shares, and timing the issue to take advantage of a rising market,
Argus could walk away with a 30 percent interest in Massey
or perhaps even majority control, and achieve this quite cheaply.

As the plan stood, Massey had little certainty that anything
would come of it. First, the plan had to be approved by common
and preferred shareholders and long-term debt holders. Second,
a proxy statement and prospectus had to be submitted to the
Securities and Exchange Commission in Washington, and the
offering had then to fall within ninety days of the latest reporting
period. But when could the company expect to lift its earnings
sufficiently to impress the market? Certainly, there seemed no
likelihood that this would be done before the summer of 1980
at the earliest.

The difficulties were increased by the fact that interest
rates were rising, adding to the financial strain on the company.
In the next few weeks, Rice and his management team were

brought before the underwriters, led by Dominion Securities and including the major Canadian investment houses, Wood Gundy, A.E. Ames, Burns Fry, and Richardson Securities, to review the company's progress to date. These meetings became regular sessions. Eventually, it was hoped, things would begin to move forward.

Conrad Black's view of Massey was that he alone would decide what would happen next. The refinancing was being initiated by Argus and it would only take place when there was a demonstrable and continuous improvement in the operating results, and when the market was ready for it. The real meetings on the refinancing plan were being held without Rice's participation; over luncheons at the Toronto Club where Black would consult with Russell Harrison of the Canadian Imperial Bank of Commerce and Tony Fell of Dominion Securities and arbitrate on Massey's future.

For Rice and his management team, it was a disconcerting experience. In their view, they had in the past fifteen months achieved what had been expected of them; management had been restructured and inefficient operations had been cut. At great cost, in human terms, Massey had been brought to a break-even position. Despite this, at the close of 1979, there was no guarantee that the company would get the equity it needed or be able to fend off bankruptcy.

15

Constructing a House of Cards

Systematically and skilfully, Rice had leveraged the company into a position where it had a reasonable chance of survival. But the exercise had been rather like building a house of cards. Any puff of wind or involuntary movement could collapse the whole thing.

At first, Victor Rice had taken a relaxed attitude to the new preferred equity issue. There was a desperate need for cash. But, on the results that Massey had reported, the underwriters would have difficulty marketing the issue. Interest rates had started moving up, signalling the unwelcome prospect of Massey's having to pay painfully high rates for its new preferred. Rice felt there was still time. By the summer of 1980, according to the experts, interest rates would be declining again. It might therefore be better to wait. Equity markets would be more receptive by then.

Rice also felt hopeful about the performance that Massey would be able to turn in. Providing there was an improvement in the first half of 1980, he would be able to point to a real turnaround, something which would convince both the financial markets and Conrad Black that Massey was worth saving. There were a number of favorable signs. In the final quarter of 1979, sales had been $1 billion for the first time ever. On the basis of this, a few brave financial analysts were predicting that the company might end up earning $2 or $3 a share in the year ahead, its first real profit in four years. These tentative predictions were being fully supported by Rice. The major part of the company's reorganization had been completed successfully. Markets were turning up strongly, and new emphasis was being put on selling into the U.S. corn belt where there were good prospects for growth. In February, Rice went before a group of financial analysts in Montreal and assured them that by 1981 the company would be earning an operating profit of $100 million.

His confidence was based on the sales reports and estimates that were coming in to head office. The nature of the farm equipment business, with its rapid fluctuations, makes all companies sensitive to what is happening and careful to monitor sales on a weekly basis. Sudden changes in the market must be relayed back to head office quickly so that decisions can be made to trim production or raise it. In the fall of the previous year, Massey had missed out. Preoccupied during the previous nine months with the need to retrench, it had failed to respond quickly enough when markets in North Amercia began to improve. But by the end of the year, it had caught up and the outlook and the mood within the company was one of optimism.

The optimism seemed to be shared by Black. There were plenty of worries still; interest rates had not eased up and the high value of the British pound continued to drive up costs. But Massey had had reasonably good first-quarter figures. And Black had declared that, as the representative of the largest share-holder, he was confident "at least a partial" equity financing could be completed during the year. Rice was buoyed by this. He was later to trace February, 1980, as the time when he was most hopeful that Black would give his support, and Argus would come through. In many ways, this period was the high watermark of Black's involvement in Massey. Unfortunately it was not to last long.

Those same sales reports that had shown a steady upward trend in the important U.S. market continued to look good in February, and into the first part of March. Then, on or about March 15, there was a sudden and dramatic collapse. The reports that came in for the rest of the month, and into April and May, showed that the market for tractors and combines had plummeted downwards. Moreover, the trend in the U.S. was catching. Other markets, in Canada and Europe, began to deteriorate.

The annual meeting had been held on the last day of February, and there had been no awareness then of the trouble ahead. Cautious and seemingly positive statements had been made. Massey would not be able to pay dividends on its common shares but, as part of the refinancing, arrears of preferred dividends would be paid off towards the end of the year. Rice claimed the company's record of the previous year was the greatest turnaround in Canadian corporate history.

Not surprisingly, when markets crashed a few weeks later, the mood changed from one of elation to one of gloom. In the U.S., President Jimmy Carter had ordered an embargo on grain exports to the Soviet Union. Oil prices had increased; the dollar was falling on foreign exchange markets, while the British pound was rising; inflation was being driven up; and, most

serious of all, interest rates continued to climb.* The spiralling
cost of money was hurting the company on all sides. Of its
total debt, the major share — some $900 million — was made
up of short-term loans which carried variable interest rates.
Every upward twist in prime lending rates was adding to the
company's interest expense, and draining away its limited
financial resources. At the same time, high interest rates and
rising fuel and fertilizer costs had made farmers draw back.

The meetings held at Massey concentrated not on the
immediate problem but on the prospects for the following
September and October, traditionally the big sales months for
farm equipment. It was hoped that a recovery would take shape
by then. Workers at International Harvester had gone on strike
and that would allow Massey to make more sales in the U.S.
And interest rates might start to come down. These hopes were
not kept alive for very long. It soon became plain that, whatever
happened in the financial markets, North American and
European farmers had taken fright. Decisions on buying new
equipment were going to be postponed as farm incomes fell.
Sales of farm machinery in the U.S. and Canada dropped by
29 percent in March and 39 percent in April. By May, even
the ebullient Rice was admitting that Massey would sustain an
operating loss in 1980.

The crisis that this created within the company was serious
enough, but the real crisis was created by Conrad Black. Despite
his continued statements that he had not lost faith in Massey,
Black chose this moment to stand down as chairman and chief
executive officer and revert to being the chairman of the executive
committee of the board. Black reached for a colorful phrase
when the decision was announced. Argus had ridden all the way
down the rollercoaster ride with Massey and would ride all the
way back up again, he told the press. Although the company
put out a supporting statement saying Black's move out of his
Massey office would enable all possible sources of refinancing to
be explored, Massey's prospects were looking bleaker. And the
public perception was that Black was growing disenchanted.

Behind the scenes, the division between Black and Rice
had been growing. To Rice, it seemed that Black regarded him
as an operating manager and nothing more. This irked Rice on
a personal level, but it also reinforced his determination to get
an out-and-out commitment that Argus would participate in a
refinancing. Publicly, he stated on several occasions that new
equity must be forthcoming by the end of October; if not,

* Setting the trend for interest rates worldwide, the prime lending rate in the U.S. —
which banks extend to their most creditworthy customers — rose from 13.5 percent in
October, 1979, to 20 percent by March 15, 1980. It then went down rapidly to
12 percent in the next three months, but was back to 20 percent by the end of the year,
and fluctuated around this level through the first half of 1981.

Massey would be in breach of debt covenants and the bankers would be able to move in and call the loans. In short, the company would face bankruptcy.

Nominally, Black continued to express his confidence in Massey and to voice his commitment. However, it was evident that he did not see its problems in the same light as he had eighteen months earlier when he had staked his business reputation on turning the company around. The chairman's office was vacant once again.

Black now devoted most of his efforts to the hybrid company which would make or break his reputation much more surely, Hollinger Argus. He had not only moved to get greater control over the more desirable Argus properties, but had also engineered an entry into the high-profile energy business through the purchase of an interest in Norcen Energy Resources. Later in the summer, he was to propose yet another restructuring to create a company that would be controlled by Ravelston and include Hollinger Argus, Norcen and Labrador Mining and Exploration. The intent of the deal was to transfer the iron ore assets of Labrador to Norcen so that its royalty income could be sheltered and used in oil and gas exploration. When he broached this deal, Black announced that the Argus stake in Massey had been written off. It was given no value, and would have no bearing, on his plans for Norcen and Hollinger Argus.

Though his stay in the chairman's office had been brief, Black contended that it had been enormously helpful. The Argus reputation had staved off the bankers and kept Massey afloat. Without his involvement, the company would have gone bust in 1978. These statements, however, did nothing to ease Massey's predicament in the spring and summer of 1980. Moreover, they raised the question of why — with another crisis imminent and help more desperately needed than ever — Black had chosen this moment to downgrade his involvement and take himself out of the company as chairman and chief executive officer.

The titles that Black had given up were taken over by Victor Rice. However, it soon became apparent that the full authority that Black had exercised was not being handed over. Rice was chairman of a company, and chairman of a board of directors, that was controlled by Argus. Representatives of the Argus board who also sat on the Massey board included the Black brothers, Dixon Chant, Fredrik Eaton, Hal Jackman, David Radler, and Trumbull Warren, and the executive committee of the board was entirely composed of Argus members with the exception of Rice himself. Even though he now had doubts about Argus's willingness to join in the refinancing, and about whether assistance from Argus was the right thing for Massey, Rice could not take any action without the approval of the directors.

Black was no longer chairman of the board. Still, he continued to exert much the same influence over Massey's future and its plans as he had done previously. As long as he kept on saying that Argus would participate in an equity issue, Rice could not seriously pursue other methods of financing or find other business partners. Meanwhile time was running out, and the financial state of Massey was becoming more perilous every week.

Discussions were held with others. Sensing that Black's retreat from the chairmanship could mean a further reassessment of his role, Rice formed an equity team within the company consisting of himself, and his senior officers, Vincent Laurenzo, Douglas Barker, Brian Long, and Don Douglass. Their job was to float proposals for refinancing in the event that Argus did agree to participate in a more limited way, or in the event that it did not.

Rice, meanwhile, attempted to resolve the situation with Black. By this time, relations between the two had reached an impasse. Rice confided to friends that he thought Black might be looking for an excuse to fire him, and that he would be dismissed for incompetence and replaced by a president who was more co-operative. When he told Black that alternative schemes for refinancing should be examined, he was informed that Argus was looking after the company's financial affairs and the matter was outside his sphere of authority. The bank with the biggest financial commitment to Massey, the Canadian Imperial Bank of Commerce, continued to deal exclusively with Conrad Black and had an agreement with him that he would stick with Massey. So Rice was effectively cut out.

Nothing could be done as long as this deadlock persisted. But attempts were made, without Argus and the board of directors, to sound out other possibilities. A U.S. investment firm was employed to look for alternatives, and partners. There were tentative discussions with the Canada Development Corporation, a government-owned corporation with a public shareholding and $3.5 billion in assets, and with Brascan, one of Canada's largest managment companies controlled by the Bronfman interests. The Alberta Heritage Fund, which had $7 billion oil money in its kitty, was also approached. The U.S.-based Caterpillar Incorporated and TIC Investment Corp. a holding company that had acquired the farm equipment interests of the bankrupt White Motor Corporation, were invited to talk about either a merger or acquisition.

The approaches were not pursued very aggressively. Nor was much hope held out. A question mark hung over Argus and its response. And Massey, now loaded down with $1.6 billion in debts against assets of only $2 billion, was scarcely an attractive buy. Any corporation, however large, would have problems consolidating such a chronic loser into its balance sheet and justifying its action to its shareholders. The most serious of the white knights, Caterpillar Incorporated, a U.S. construction machinery giant and a highly profitable company with sales of $8.6 billion, balked at the idea of taking on Massey's debt load.

As Massey executives cast the net for partners, they inevitably found themselves the victims of new rumors and speculation. Again unfavorable publicity worked against them. To buy from Massey was foolhardy, farmers were being told; the company would not be in business long enough to honor its commitments.

To keep going, the company had been forced to extend its customary month-long shutdown of North American plants to three months, laying off 5,000 workers in Brantford and Toronto, another 1,200 in Detroit and 600 in Des Moines. There was no hiding the fact that the company was sliding towards the brink. Massey shares had hit a low of $7.50. And, as the news continued to worsen, the company found itself being likened to another desperate corporate giant, the Chrysler motor car company. Rice protested at this. Massey was much stronger and better established as a global company and it had valuable properties — Perkins diesel engines, its small-tractor business, its worldwide dealer network. Still, the pressures were enough to dampen even Victor Rice's enthusiasm. "Somewhere along the way wouldn't it have been nice to have something go your way?" he told a financial journalist who interviewed him in the last week of June, 1980.*

An exercise in morale boosting.

In the middle of the summer, a critical time for Massey's fortunes, the top executives of the company left the high-pressure environs of the city for a week of meetings at an Ontario country hotel, the Millcroft Inn. Renowed for its elegance and fine cuisine, the inn had been chosen as the location at which Massey would attempt to remake itself for the future. After eighteen months of intensive planning, the company was going to devise a corporate strategy for the 1980s and beyond.

Forward planning had never been much practised at Massey.

---

* The comment was made to Barry Estabrook, reported in the *Financial Times of Canada*, July 7, 1980.

The company had grown spectacularly in the past, but it had   251
done so by a process that, in good times, could best be described
as successful opportunism. Sometimes this had worked but just
as often, in the bad times that frequently afflicted the industry,
it had led to woeful mistakes and losses. Rice was determined to
change this method of operating. He had set up a formal planning
group, appointed a director of strategic planning, and insisted
that contact should be made with the firm's divisions worldwide
and their advice sought.

The result of this effort was that the fifteen executives who
met at the Millcroft Inn had a document before them which
reviewed the company's strengths and weaknesses; examined
critically the options for the next ten years; analyzed fourteen
competing firms and the probable direction they would take;
and presented a series of possible corporate objectives and
"alternative strategy sets" from which the executives could select
the best overall strategic plan.

The exercise seemed, even to many of those taking part, a
rather fanciful one — a product of Victor Rice's optimism. For
a firm that was struggling to stay alive and suffering from fresh
disasters almost daily, it was bizarre in the extreme to spend
five days talking about what Massey would be doing in 1990.

The thrust of the discussion as it developed was on setting
up so-called straw men and destroying them. Was Massey a
manufacturer or a marketer of products? If its strength for the
future was in marketing, then how did it also see its future in
manufacturing? Was it necessary, as had been done in the past,
to own factories outright or could they be shared with others?
Given the overcapacity in the farm equipment business, would
it not be better to share manufacturing facilities? In the same
way as, for example, European automakers like Peugeot,
Renault, and Volvo had combined forces in Europe to build an
engine plant.

The outcome of the session was an agreement to put new
stress on certain aspects of the company's business. The possibility
of co-operating with other firms on major capital projects was
one of these. Others were that Massey would continue to
concentrate on farm machinery; launch itself more aggressively
into the large-tractor and combine market in North America;
attempt to maintain its top rank in the European market; and
come up with new diversification plans when it had strengthened
its financial position, which it was hoped would occur by
the mid-1980s. The company also pledged to spend more of
its resources on its staff, and on improving career opportunities
among its 47,000 employees.

In some ways, the Millcroft meeting *was* a piece of fantasy,
and an irrelevance. It could also be seen as an exercise in

morale boosting, an attempt to raise the spirits of a group of executives who needed desperately to feel that the efforts they were making were not in vain.

What was at stake was the issue of Massey's survival.

Nonetheless, the strategic plan began to take on great significance as Massey moved on to the next stage of its battle for survival. Its creditors and the governments that it was being forced to approach for help were presented with a comprehensive and well-designed plan for the future. Politicians and civil servants have a special affection for neatly packaged blueprints of the future. And Massey had provided itself with a fine one that promised the company would grow rationally and determinedly, ultimately achieving a laudable set of goals; it was a hard vision to destroy.

Over the next few weeks the strategic plan began to play a direct part in Rice's effort to save the company. The target was still $500 million in equity financing. Rice declared that government support — which Chrysler had received in the spring — would be a last resort. It was still conceivable that, with its plan, the company might yet persuade the financiers and the bankers, and the investment community that it was worthwhile investing for the long term and brighter future that lay ahead. Incentives would be required, of course, and they would have to be at the expense of the existing shareholders since a massive dilution of shares would be required. Still, shareholders were conditioned to bad news, and to the possibility that they might lose everything through bankruptcy.

The new plan involved four main elements. First, Black had to stand firm with a commitment from Argus. Second, the Canadian Imperial Bank of Commerce, which was owed $360 million, had to subscribe to the stock issue. That would yield $250 million. For the third element, Rice had in mind that a private placement could be made and an investor — confident about Massey's future potential — would come forward, while as the fourth element, Massey's long-suffering shareholders would be asked to participate in a rights issue as a do-or-die effort to save their company.

Time was running out. There could be no more waiting for an improvement in the company's performance or evidence of a turnaround. In addition, the banks would have to hold off and waive the debt covenants that were coming due, since it was doubtful that the package could be put in place quickly enough.

For a few weeks, some hope was held out. Black co-operated
by going on record in support of the scheme, although he was
anxious to make clear that Hollinger Argus would make its
investment only when everything was finalized and only because
Massey merited it, not because of any prior obligations.
Ultimately, however, the plan was to fail, primarily because
Massey's finances were in such dire straits that none of the
parties wanted to be the first to make a commitment. All of
them hung back, waiting for some move that would signal a
breakthrough had been made. Rice needed a firm pledge from
someone, but it never came.

Faced with another deadlock, Rice then began to con-
template involving government. If the stock issues could be
backed by a government guarantee, then a solution might be in
sight. Under these circumstances the fence-sitters would be willing
to make a decision. They would have nothing to lose, and
they might be persuaded that they had a lot to gain. After
all, they would be increasing their stake in Massey, and doing it
with less risk to themselves.

Approaches had already been made to the federal government in Ottawa and the provincial government of Ontario.
Overtures had also been made to the British government of
Margaret Thatcher. The company had a third of its assets and
a third of its work force in Britain, more than in any other
country, and — because of the importance of Perkins and the
Banner Lane tractor plant in Coventry — it had traditionally
been thought of as a British-based company.

Rice and the firm's chief in Britain, Michael Bird, lobbied
the British cabinet for grants under its industrial development
program. The timing of the request was not good. The Thatcher
government was attempting to cut expenditures and slash
subsidy programs, and it was faced with a gigantic money-
loser, British Leyland, which it had to support; it did not want
to underwrite Massey and Perkins as well. A strongly worded
refusal came from the leading free enterpriser in the cabinet,
Sir Keith Joseph. But this was tempered by a final paragraph
in his letter — added by Margaret Thatcher — which held out
some hope of assistance in the future.

In Canada, Black had written letters to federal Industry
Minister Herb Gray, and Rice had talked with him. At first,
the approaches were purely a precaution. Chrysler had been
petitioning the Canadian government, as well as the Carter
administration in the U.S. Massey felt that it had a stronger
claim on Ottawa, as a company with a Canadian heritage, and
Rice wanted to tell the government that Massey, too, might
need help if its efforts to get private funds failed.

By August, it was plain that other options had been

exhausted, and talks were opened with the federal government. In the past, Ottawa had shown itself prepared to buy up companies that were in trouble. In the mid-1970s, to save Canadian jobs, it had acquired two aerospace manufacturers, Canadair of Montreal and de Havilland Aircraft of Canada of Toronto. The government of Pierre Trudeau had shown few qualms about intervening in the private sector. At the time of the Massey approach, the cabinet was drafting an energy program that would increase the powers of the publicly owned oil company, Petro-Canada. A few months earlier, the government had provided $200 million in loan guarantees to bail out Chrysler Canada, while two years previously, it had invested a small sum in a farm implement company, Canadian Co-operative Implements of Winnipeg, to keep it afloat.

But if the precedents were helpful, the attitude of the Liberal government to Massey was not. In the past, the company had cultivated strong links with the Liberals, Canada's traditional ruling party. The founding Massey family had been associated with the Liberals, Massey's former president, James Duncan, had worked for Liberal strongman C.D. Howe during the war, and Eric Phillips of Argus had had a long-standing friendship with former prime minister Lester Pearson. In the 1970s this cordiality had given way to something approaching hostility. Bud McDougald of Argus detested the Liberals and branded Trudeau as a die-hard Socialist, if not a Communist. The company was remembered in Ottawa for its 1971 threat to pull out of Canada. By its own inclinations and actions, the company had chosen to downplay its Canadian roots; less than 7,000 of its 47,000 workers were located in Canada compared with 17,000 in Britain.

In the final week of August, the Trudeau cabinet considered giving financial help to Massey and putting itself forward as a backer of part of a new share issue, but then shelved the proposal for the time being. Several cabinet members had argued against it. They did not want the government to buy shares itself, nor did they want any guarantees to be offered. Their opposition stemmed not only from Massey's lack of a strong Canadian presence, but also from the commitments that seemed to have been given by the banks in Canada and abroad, who had kept supporting Massey, and by Argus. How could the government put taxpayers' money at risk to bail out foreign lenders? And how could the government be seen to be coming to the aid of Conrad Black, the prince of Canadian tycoons?

The negative reaction in Ottawa had been anticipated within Massey, and at first no great efforts were made to overcome it. Rice concentrated on talking to the lenders and attempting to

get them to ease their deadlines and roll over the debt that
was coming due. Besides its single largest debt to the Canadian Imperial Bank of Commerce, the company owed $950 million to U.S. lenders, with the largest commitments being to the Continental Illinois Bank and Citibank; it owed $450 million to British banks, led by Barclays Bank, Lloyds, and the Midland Bank; and $175 million to French banks. The remainder was distributed among Swiss, Italian, German, Australian, and South American banks and two other Canadian banks, the Toronto Dominion Bank and the Royal Bank. Rice had been frozen out from direct dealings with the Canadian Imperial Bank of Commerce by Black, but he had deliberately and carefully cultivated relations with the foreign banks, keeping them informed about Massey's plans.

While Ottawa's cool response was treated calmly by the Massey management, it touched off a storm of protest from Black. He agreed with the politicians in Ottawa that it was intolerable for the government to be asked to step in and save a group of international banks. They had lent money to Massey and continued lending, on the basis of its stature and reputation as a multinational company. Indeed, they had put up money even when the company had accumulated a huge and burdensome amount of debt. (Massey now had $4.26 worth of debt for every $1 of equity, compared to figures of $1.40 and $1.10 for International Harvester and John Deere respectively.) Why should the banks now be saved from the folly of their actions? Black had taken a strong line with the banks himself. It was a question of whose interests should come first, the shareholders or the lenders. Argus was prepared to put money into Massey if the lenders relaxed their loan terms. The banks, on the other hand, wanted to see new equity before they would offer any concessions. To this deadlock was now added Black's dispute with the government. He resented the idea that Argus was going cap-in-hand to the federal and provincial governments because of its association with Massey. And he was outraged by the way this was being treated by the media: "Why should the taxpayer bail out the fat cats of Argus?" inquired one newspaper headline.

One characteristic of Black was his strong and personal reaction to any form of public criticism. Since the days when he had stormed around newspaper offices demanding that they retract unfavorable reviews of his book on Duplessis, he had not had to worry much. The media had been respectful, and full of praise. Now the troubles of Massey were bringing him unfavorable reviews again. His first reaction was to disavow his obligation to Massey. He was not responsible for creating

Massey's difficulties, and it was unfair that he, personally, should be criticised. His second reaction was to put a greater distance between himself and the company, and to devise his own arguments as to why Massey was seeking government aid and support.

To Black, the help that was being sought was to strengthen an old and valued Canadian firm. Massey ranked as the largest industrial company in the country in terms of its sales. The company had a manufacturing base and a position in important markets around the world that was unique; it could be a source of new jobs and new investments that would benefit Canadians, so why should the government be reluctant to become involved?

Black's letters to Herb Gray were along these lines and referred to the industrial benefits an investment in Massey might bring. He also emphasized that allegations that Argus was seeking a government bail out for itself were irrelevant and unfair. Black's motives in engaging in this correspondence remain hard to fathom; later he came to regard these private pleas as a matter for public comment, and the letter were passed along to a reporter from the *Toronto Star*, Irv Lutsky, for publication. There was an obvious element of self-justification involved. Having rationalized, four months earlier, that Massey would be better served by him stepping down as chairman, he was now rationalizing that Argus had become an obstacle to government participation. Would it not therefore be better for Argus to remove itself from Massey altogether?

To Rice and the other Massey executives, Black's arguments made little sense. If the governments in Ottawa and Ontario were to come to Massey's rescue, it was hard to see what immediate benefits would flow to Canada. Obviously, some trade offs could be negotiated. Providing Massey staged a recovery, it could agree to expand its diesel engine facilities and make new investments and locate them in Canada rather than abroad. But the governments were not needed as providers of new funds in order for Massey to enlarge itself and create new jobs; they were needed to break the impasse. Argus might be willing to participate in an equity issue, but it was obviously not prepared to launch one. To get the money it required, the company needed someone to take the initiative at the eleventh hour, providing a vote of confidence and a cast-iron guarantee.

The question of government involvement, however, was less important than the question of what Argus would do next. Rice's equity team was by now taking a jaundiced view of the way things were developing. The Argus-led board of directors had never worked well. Few of them had any knowledge of the industry or the company, and they would raise picayune little issues about the state of the combine market in Saskatchewan or tractor sales in Sweden while avoiding the explosive issue of the refinancing. Nominally, Argus would participate in the refinancing, but when and how was never decided.

In this atmosphere, a number of executives began to plot how the main impediment to their plans, Argus, could be removed. It was wishful thinking, but it sprang from the fact that they were now doubtful that Argus would give Massey any support at all. These doubts were reinforced by the way Black was treating the idea of government help. He was looking at what it was going to do to the reputation of Argus and himself, not at Massey's near-desperate need to resolve its problems. Moreover, he seemed to take the attitude that the foreign banks must be leaned on more heavily, and wrote a letter to Rice to this effect. To Black, the responsibility for bailing out Massey was a shared one. Others had to make a move before Argus would.

On October 1, Rice flew to Des Moines to attend the biennial U.S. Farm Progress Show which was to be staged in rural Iowa, about 150 miles from the city, the following day. Since he could not make airline connections quickly or conveniently, and Massey had sold its own corporate aircraft, he went in the Argus plane. That night, he was the guest of honor at a dinner attended by members of Massey's dealer council in the U.S. In his speech to the dealers he admitted the company was in bad trouble, but, he said, with their help and support it would survive. On the following morning, he woke up early and was driven to the show. Once there, he made his way to the Massey exhibit to shake hands and greet the dealers and farmers. He had been there just five minutes when he was told that a telephone message had been received from Conrad Black. The telephone had been set up at the back of a trailer, and Rice went there and phoned back to Toronto.

He was on the phone for the next five hours. Black did not tell him directly that Argus was pulling out of Massey. He said that he and his friends had discussed giving the shares away and he intended to consult with them further. In the space of the next few hours, his position changed. At first he had been considering the matter, but, by the end of the afternoon, Black was firm. The Argus directors on the board would send in letters of resignation, and the largest shareholding in Massey would be given away to the firm's pension fund. On the previous day, the company's vice-president of employee relations, Phil Moate, had received a request from Argus to send around copies of the pension fund agreements. So Black had apparently been deciding on a course of action then.*

The gesture was a dramatic one. There were other alternatives Black might have taken: For example, he could have kept the shareholding and surrendered control of the board, thereby allowing management to act independently of Argus. The gift to the pension fund put Rice and his management team clearly in charge. They were responsible for administering the fund themselves. Unfortunately, it also signalled that Argus was pulling out and wanted nothing more to do with Massey.

Rice contacted Laurenzo and his senior staff in Toronto to tell them the latest development, and then drove back to Des Moines and flew to Ottawa aboard the Argus plane. He had managed to arrange an evening meeting with Industry Minister Herb Gray, and later phoned Bill Davis, the premier of Ontario, from his hotel. His first thought was that the Ottawa and Ontario governments had to be told personally that Argus was abandoning Massey, before the news was published in the papers the following day.

The walk away from Massey.

The decision by Conrad Black to walk away from Massey was as surprising and controversial as any business decision ever made in Canada. Black's justification for it was that Argus had been put in an impossible position; the attitude of the government was that any assistance to Massey would be seen as bailing out Argus. Far better, then, for Argus to remove itself from the scene. Both Massey and the government would be in a position to go forward with their plans.

However, other parties in the Massey affair saw no such compulsion, nor had the negotiations reached a stage where

---

* Besides getting away from the tangled financial affairs of Massey, Black also earned for Argus a capital loss to carry forward against future capital gains. He had disposed of an investment that had, for all intents and purposes been written off, but was given a $27.8 million valuation on the Argus books. His gift to the pension funds put Massey effectively in the hands of Rice and his executive team who were charged with administering the funds.

pressure was being put on Black to reach any such monumental 259 decision. Federal officials were flabbergasted when they heard the news. They had been working all along on the premise that Argus would be involved. Black's departure raised new and disturbing questions about whether the company was worth helping.

In the past, Black had gone on record as a forthright opponent of government involvement in the private sector. He was plainly uncomfortable with the role that he was being forced to play. He had other fields to conquer and was transforming the business empire that he had won into something more in tune with the times, and more profitable. Perhaps Massey could have been fitted into these plans if it had made a quick and creditable recovery. But it had not. And it was apparent that, even if markets turned around after 1981, Argus as the principal shareholder would have to carry on fulfilling its obligations for years to come — and doing so in the public spotlight, with governments and the media looking on and questioning every move.

Certainly Black's argument that he personally did not have any debts to the past was a thin one. Having fought so determinedly to gain control of Bud McDougald's inheritance, he might have felt more strongly disposed to accept the legacy that went with it. Massey was, after all, weak and strapped for cash because Argus had neglected it for many years. The representatives of Argus had put themselves into a position of hegemony over the company. They alone could pass judgment on Massey's financial needs and supervise its management, and they had failed to exercise their mandate.

Moreover, for a self-appointed spokesman for free enterprise like Black to withdraw from Massey was to admit that something very fundamental had gone wrong. He was, after all, declaring that a great Canadian company — with 133 years of history behind it — would be better served by putting itself in the hands of governments and banks than by enjoying the continued participation of its largest private shareholder. Tactically, he may have acted in the best interests of himself and his friends who ran Hollinger Argus. But the precedent of admitting defeat, and leaving the problems to be solved by others, was hardly a helpful one.

Certainly, it was of no immediate help to Rice. One creditor, Dresdner Bank of West Germany, was threatening to call a $10 million loan. With Argus's withdrawal, other banks might follow. Moreover, the governments in Ottawa and Ontario were in the first stage of considering aid to Massey and were likely to move with bureaucratic slowness towards a final decision.

In Ottawa, a meeting of the federal economic development committee drew only a cautious statement calling on the Canadian Imperial Bank of Commerce and Massey to work together on a refinancing plan. In Ontario, Industry Minister Larry Grossman would only say that the government had not ruled out aid for Massey. Three weeks later, the governments came up with a qualified commitment. They were "prepared to guarantee the capital risk of a portion of new equity investment" from the private sector "providing various conditions are met including a satisfactory degree of co-operation from existing lenders." Ottawa selected two representatives of its own, John Abell, vice-chairman of Wood Gundy, and Gordon Lackenbawer of Pitfield Mackay Ross, to work on developing a refinancing plan.

In the immediate aftermath of Black's withdrawal, Victor Rice had reacted in two ways. At first he was concerned that the abandonment by Argus would prove to be Massey's death blow. The government would back down. The banks would become agitated and the financial community would conclude that, if Argus was not prepared to stay the course, no other private investors would be enticed to put money in. His second response was one of confidence, and some elation. He had sat around the table with his senior executives debating Argus's next move and deploring the way Black was obstructing what needed to be done. Now, the obstacle had been removed. Providing the bad news of the Argus departure could be contained, then the company had the opportunity to save itself.

The first sign of the changed relationship was a phone call from Russell Harrison of the Canadian Imperial Bank of Commerce. The bank had always dealt with Black before. Now Harrison was telling Rice that the CIBC was the only thing standing between Massey and bankruptcy, and they would like to be more actively involved. Seven Argus directors had left the board. Charles Laidley, a vice-chairman of the bank was brought on to it.

During the first two weeks of October, Rice moved quickly. He did so both to provide reassurance that Massey was not on the point of collapse, and because he had detailed financing plans to work with. His equity team had been laboring for months assessing the alternatives if Argus should pull out. One day after journeying from Des Moines to Ottawa, he was back in Toronto discussing the plans with his staff. Three days later, he called a meeting of the depleted board of directors and presented them with the proposals.

The financial rescue package as it emerged, and was made public, involved the injection of $700 million in new preferred and common shares. This would consist of a government guaranteed portion of $200 million in common or preferred shares; $150 million in convertible preferred to be bought by the Canadian Imperial Bank of Commerce; and $350 million in convertible preferred, to be issued to the public, guaranteed by the other banks, with the Commerce purchasing any unsold amount up to $150 million. If the package won approval from the lenders and shareholders, then it would mean the infusion of $425 million in new equity and the conversion of $275 million of debt into equity, a process that would greatly improve the company's balance sheet.*

Since both Ottawa and Ontario gave their support in principle, the next stage was for Rice to tackle the foreign banks and gain their agreement. By the end of October, he had launched himself on an exhaustive schedule of visits to bankers in Europe and North America. These meetings were to culminate in an all-lenders conference in London's Dorchester Hotel in December.

In the state that Massey was in, the arguments for saving the company were as much negative as positive. For the share-holders the alternative was to agree to seeing the value of their shares watered down drastically or lose their investment altogether. For the banks the choice was to reach a deal or engage in the messy business of rendering Massey insolvent. Almost none of the lenders had any kind of security on their loans. For the governments the choice was to safeguard over 6,000 jobs in Canada and keep an important industry intact, or risk the political flak that would have come from Massey's bankruptcy. The province of Ontario, the location of Massey's factories in Canada, had a particular motive for helping; the Progressive Conservative government was on the verge of facing an election in which the crucial issue would be the state of the economy and the trouble facing Ontario's manufacturing industry.

As it turned out, the British government also felt it had something to lose — and nothing to gain — from Massey's demise. The Thatcher government stayed clear of any form of direct assistance, but the Bank of England put pressure on the British banks, led by Barclays, to go along with Massey's plans and take equity in the company. Sir Richard Pease, vice-chairman of Barclays, was an early convert to the refinancing —

---

* The formal rescue package could also be seen as an informal bankruptcy since its effect would be the same; shareholders would see their investment diluted to a fraction of its worth while the creditors would give up their preferred position and take an ownership position.

which was more vital to Britain than to any other country since Massey's work force there was nearly three times larger than in the next biggest countries, Canada and Brazil.* Barclays ranked as Massey's second largest lender after the Commerce, so its support was crucial. But Pease contended that Barclays would be better served by taking equity only in the British operations and ignoring the rest.

For Rice the sheer diversity of Massey's debts, which were scattered around the world, was a problem in itself. He and a negotiating team, as well as John Abell and Charles Laidley, concentrated on the major U.S. and British banks. If the main rationale for the lenders and governments was to avoid something worse, the more complicated situation of Massey going bankrupt, this was not the case for Rice. He was convinced that the company could prosper if its huge debt load could be lifted for a sufficient period of time and he devoted a lot of time to trying to persuade others of Massey's prospects.

Rice also had to fight off unexpected opposition. The retired chief of Perkins, Sir Monty Prichard, appeared on the scene with proposals that would have scuttled the whole plan. Prichard had long set his sights on controlling Perkins. Now he proposed to the governments in both Britain and Canada that Massey should be dismembered. The Canadian company could be run out of plants in Toronto and Brantford, while he would take charge of Perkins in Peterborough and the tractor plant at Coventry. It was an appealing idea in both London and Ottawa since both companies would be viable and jobs would be protected. But it would have meant the end of Massey. Rice, angered by the interference of his former boss, had to scotch the idea quickly.

Another obstacle came from the U.S. banks which were reluctant to guarantee their part of the equity financing. In the middle of November, a group of bankers led by Continental Illinois declared that they did not favor the scheme. They would be prepared to help, but they would rather do so by forgiving interest on outstanding loans. The U.S. banks had to be listened to since they held a strong hand. In the event of bankruptcy, they could gain easy access to the richest pool of Massey assets.

---

* In terms of numbers of employees, and the stake each country had in Massey's survival, the figures were: Britain, 17,000; Canada, 6,700; Brazil, 6,000; the United States, 5,500; France, 4,900; Italy, 3,200; West Germany, 1,900; Australia, 1,400; and Argentina, 400. Besides Canada and Britain, an attempt was also made to interest the French government in assisting Massey. Approaches were made shortly after Argus withdrew but they were unsuccessful.

The culmination of the refinancing effort came at three meetings; one with the shareholders in Toronto, and two with the bankers in London. The success of these meetings was crucial. The meeting with the shareholders was the easier one for Rice. About 500 shareholders attended. The meeting lasted for four hours as common and preferred shareholders tried to get a definitive statement from Rice on how the rescue plan was going, and how their holdings in the company would suffer. Rice would only tell them that the plan was half-completed; that the company was managing to survive by paying salary and interest charges out of its retail cash flow; and that there would be a substantial dilution of shares. The alternative Rice told them, would be to lose everything. Little more information could be divulged without the company laying itself open to legal suits and being held in violation of securities laws. Some bitterness came to the surface. Shareholders were critical of the continued presence of directors such as Albert Thornbrough on the board. But little of the acrimony was directed at Rice and his management team. The shareholders endorsed the refinancing, and the creation of new preferred and common shares.

Ahead of the London meetings, Massey had released more bad news. The company had come through a disastrous final quarter of the year that had pushed its total loss for the year to $225 million, and its operating loss to $139 million (actually a worse figure than in 1978). In the last three months of the year, sales had declined by 32 percent, and there were no prospects for improvement. In the first quarter of 1981, the company would lose another $100 million. The fresh disaster caused some banks to get nervous. An Australian creditor broke ranks and demanded payment of a $4 million loan.* Meanwhile the Bank of Commerce announced that it had set aside $100 million for possible losses associated with its loans to Massey.

The doleful news created some doubt about the success of Massey's next effort, two week-long meetings that took place in London's Dorchester Hotel in December and January, which was intended to bring together all the worldwide lenders to

---

* When Massey refused to honor bills of exchange called by Capel Court Corporation, an Australian lender, the firm initiated court proceedings. Rather than break the agreement with other lenders — placing a moratorium on debt repayment — Massey-Ferguson Holdings (Australia) placed its operations in the hands of a receiver, thereby deferring the debt repayments for the time being.

264 hammer out a final agreement. Systematically and skilfully, Rice had leveraged the company into a position whereby it had a reasonable chance of survival. But the exercise had been rather like building a house of cards. Any puff of wind or involuntary movement could collapse the whole thing.

The old problem that had existed with Conrad Black and Argus remained. No one party would step forward and be the first to make a commitment. Since neither the governments nor the lenders nor the shareholders wanted Massey to go under, all of them were prepared to co-operate providing the others did. The shareholders had already coalesced in this, and had nothing to lose by it. But Canadian government support was conditional on "a satisfactory degree of co-operation from existing lenders;" the Bank of Commerce would go along only if other banks did; and the foreign banks would sign nothing unless it involved the participating governments, and would agree to nothing if they felt it put them at any disadvantage with other banks. By far the most complex situation was that presented by the banks, coming as they did from different countries with different rules and practices, and varying in size and in their degree of financial commitment to Massey.

There were 250 banks whose representatives had to be individually reassured. And, although they had turned a blind eye to the company being in default on its agreed debt ratios, the spirit of co-operation had not extended to the point where the company was being allowed to operate normally. This was the heart of Massey's difficulties: Because of the actions of the banks, the company faced a severe liquidity squeeze.

In each of the countries concerned, the banks had lent money to Massey's operating subsidiaries. If the company had been healthy, then it would have been able to use funds from its subsidiaries in the form of intracompany transfers. However, as soon as its troubles became known, the lenders started imposing restrictions on this flow of funds. They did not want to lose control over assets they might be able to obtain in the event of a bankruptcy. This left Massey with a severe shortage of working capital, and tough cash conservation measures had to be imposed.

In negotiating with the bankers in London, Massey was looking for an agreement in principle on its refinancing plans. But it could not hope or expect that it would get much relief or improvement in its working capital position. Until an actual agreement was signed by all the banks — something that would come much later — Massey would continue to have money problems which would set back its overall performance and put its survival in jeopardy.

For the week of meetings at the Dorchester in January, fifty rooms and ten suites were taken by the company and its bankers. Rice and Massey's treasurer, Douglas Barker, talked to the bankers in their suites and toured the hotel corridors, reassuring creditors that no secret advantages were being gained and that Massey would continue to be fully involved in their country. The Europeans, from Britain, France, Germany, and Italy, were keen to extract promises that Massey would keep up its investment in their countries. At the Dorchester, the ballroom had been turned into an international forum, with simultaneous translation facilities by teams of interpreters. After each round of individual meetings, Rice would call a full session to reassure all the bankers, large and small, that they were being treated equally.

Complete unanimity was needed. A pull-out by even one bank would have overturned the agreement with all the rest. The representative of Barclays Bank, for example, kept pressing the idea that Massey's operations in Britain would do much better if they were separated from the Canadian company. The last session was due to start in the ballroom at 6:00 P.M. on Friday, January 16, but it was delayed for ninety minutes as Rice consulted with European bankers on the final agreement. When he emerged from that meeting, he told the bankers: "Gentlemen, we have a deal." The news was greeted with some applause. The bankers were more relieved than excited that the long-running saga of Massey's financial troubles had reached some kind of conclusion.

During the week, there had been plenty of compromises. The most significant occurred when the Europeans had been out-manoeuvred by the Americans and had agreed to go along with an interest forgiveness scheme. According to the equal-treatment-for-all formula, each of the banks, regardless of the size of its loan or the rate of interest, would be able to convert into shares interest amounting to 22.5 percent of what had been lent. The total amount of interest forgiven would be $280 million and the banks would take thirty-seven million shares. In addition, Massey would be able to raise $90 million from investors on the sale of a preferred issue guaranteed by Britain's Export Credit Guarantee Department. It could also call on $100 million in debt conversion and $50 million in new equity from the Bank of Commerce.

The results of the Dorchester meeting had been impressive. In effect, the banks have given the company most of what it wanted and needed. For Rice, who had spent many weeks cajoling and pleading with the bankers, the outcome was especially rewarding. The Ottawa and Ontario governments had said they would do nothing unless the lenders came forward with a rescue effort. Now he would be able to present them with a fait accompli — the banks were willing to do their part.

On paper, things looked good for the company, and it had

apparently pulled off a remarkable transformation. In the immediate aftermath of the bankers' meetings, spirits were high and it seemed likely that Massey would be able to piece together the remaining parts of the puzzle quickly. Essentially, all that remained was to negotiate satisfactory terms with the governments in Ottawa and Ontario, and proceed with an issue of preferred shares, which would have their support.

As it turned out, Massey's problems did not go away so easily or so quickly. In part, this was because the process of bargaining with the governments and then finding backers took a great deal of time. But mainly it was because the company's relations with its bankers were so excessively complicated, and the legal implications of what was being done were unprecedented. As a group, the banks might have agreed to a relaxation. They would commit themselves to taking shares and forgiving interest up to a certain ceiling and to restructuring debt, but this did not prevent them from continuing to restrict the transfer of funds, or continuing to be edgy about the commitments they had made. The banks were restive and they became uneasy and unhelpful as the delays went on, and no final agreement was reached. The date for actually signing a legally binding document continued to be put back. Meanwhile some banks threatened to change their minds and break ranks.

In February, Rice had received the government guarantees, and the job of placing the preferred issue with banks and financial institutions was handed over to the company's financial advisers, A.E. Ames and Company. The main undertaking given to the governments was that Massey would maintain jobs in Canada and establish a Canadian research and engineering base.

Despite the agreement, the talks did not go smoothly. Ottawa was still reluctant, and anxious to attach as many conditions as possible to its support. The provincial government in Ontario was receptive since it faced an election the following month and wanted to demonstrate its support for Massey. Both governments were now faced with having to make a full guarantee.

After the federal cabinet divided on the issue, a final decision was made by Prime Minister Pierre Trudeau.* Ontario raised its stake to $75 million, and Ottawa lowered its participation to $125 million — on the basis that a payout would occur if Massey failed to meet dividend payments on the shares.

---

* Talks had also taken place with two potential rescuers of Massey who had close links with Trudeau and the Liberal government in Ottawa. Maurice Strong, a former chairman of the national oil company PetroCanada and a man with an interest in Third World affairs, and Ottawa developer Robert Campeau had both been identified as so-called white knights. As it turned out, neither was involved in the rescue, but Strong, a former United Nations official and chief of Canada's foreign aid agency, remains a likely candidate for Massey's board of directors, due to be enlarged in 1982.

Following the agreement, Ames' president Robert Bellamy said he was confident that the issue could be placed within ten days. However, this prediction proved to be an embarrassment. It was not until the end of April, after the terms had been revised and the dividend increased, that the placement was made, with Canada's largest bank, the Royal Bank, agreeing to take a $100 million block.

The delays and revisions were costly for Massey. The company had to arrange interim financing from its suppliers as it waited on the refinancing package. The drawn-out negotiations and concessions illustrated how tough the battle had been and how hard it was to create a climate of confidence again. Massey continued to chalk up phenomenal losses. It had managed to get its rescue plan approved, but the victory had been a narrow one. It was not until July that an army of bankers and lawyers finally came together to sign a comprehensive refinancing document. Victor Rice had shown himself capable of pruning and cutting back and making Massey's operations more efficient, but he had not shown that the company could be made profitable. The risks were still there, and the bankers were very aware of them. The refinancing would help pull the company back from the brink, but it could still face failure and bankruptcy.

16

Looking Forward and Looking Back

While Massey's successes and failures can be traced to its management, the personalities that have run it, the limitations of the Canadian market and the vigor with which new policies and products have been implemented, the single greatest influence has been the boom-and-bust cycle of farm markets around the world.

In any company that has passed through as much history as Massey, there is a sense of shared experiences and a loyalty to the past. Often, this feeling for tradition has been put to the test. A new management has arrived on the scene, determined to change things, and to correct the errors of judgment and bad habits of its predecessors, but in time the sense of continuity and of a shared tradition begins to reassert itself again.

Shortly after Victor Rice announced a breakthrough had been achieved on the refinancing, he received a congratulatory letter from the last person to hold the post of chairman and president of the company, James Duncan. Now living in Bermuda and well into his eighties, Duncan wrote to Rice as a gesture of good will. But he also thought it would be helpful to set down, in great detail, the troubles that had afflicted Massey as it lost money one year after another during the Depression.

At the time Duncan had been credited with being the savior of the company, for he had stood firm, much as Rice had done. He had refused to be panicked into dismembering the company and kept alive the idea of Massey as a global enterprise. He had also managed to enlist the support of the company's bankers to keep Massey from sinking under the weight of its debt. Rice had been called on to respond to the same challenges nearly forty years later.

The parallels between the past and the present have recurred time and again. Some have concerned important issues of principle and management strategy, while others have been on a purely personal level and relatively minor.

One of the laws that has governed Massey's operations from generation to generation is that the company has done well when its presidents have been young, vigorous, and fresh with new ideas. Later, when the same presidents have become accustomed to the exercise of power and complacent in using it, the company has invariably run into problems.* Another lesson from the past is that the company has had to pay dearly whenever it has fallen behind in the development of new products, and in the marketing of them.

In the 1920s, Massey failed to move aggressively into the new market for tractors. As a result, the company limped along, doing relatively poorly, until it finally solved the problem by an arranged merger with the Ferguson company. Similarly, in the 1970s, Massey pursued other lines of business, confident that it could establish itself as a supplier of industrial and construction machinery, and neglected its farm products, the mainstay of its business. This became apparent in two markets, both of them crucial for the company's success — large tractors and combine harvesters. When it failed to stay abreast of its competitors, sales began to tumble, particularly as farmers became more selective in their buying.

In marketing, the problems of the past have surfaced again. And so have the solutions. When John Ruth was appointed to head worldwide marketing operations in 1979, he started with a clean slate. Massey had decentralized itself to the point where it had lost control of its sales network, and was not responding to what was happening in the marketplace. It was only when Ruth began to impose new rules and marketing codes that he realized that there was a precedent for what he was attempting. In the early years of Thornbrough's presidency, when the emphasis had been on new ideas and approaches, a Texan named John Shiner had been given the same mission.

Like many international companies, Massey has had to formulate and reformulate its basic management structure. There has always been a natural tension between the head office in Toronto and the operating units spread out across the world, and it has been difficult to achieve a balance. Too much decentralization led to the excesses of the final Thornbrough years. But too much control and authority at the centre has

* Curious parallels can be drawn. The company archives at Massey contain a film made of former president James Duncan's last world tour, showing him being greeted like a visiting head of state. Duncan loved the trappings of office and made full use of the prestige associated with the Massey name. Thornbrough, when he assumed the presidency, was much less concerned with such things. But, in the 1970s, he, too, developed a taste for making regal visitations. Aware of the precedents, Victor Rice has kept things simple to date. Just forty years old, he says he will not keep his present job for long and has already decided on a date when he will stop being chairman, president and chief executive officer of the company.

proved just as disruptive. Always the problem has been how to combine local initiative with centralized control and responsiveness. And it has not often been satisfactorily resolved.

In some ways, Massey's latest crisis has been different from previous ones. Certainly it has been the most serious. But, in an industrial enterprise with origins dating back to the 1840s and an earlier industrial age, similar crises have been played and replayed before. Almost as soon as the Canadian west was developed, Massey found it had outgrown its manufacturing base in its own country. As an industrial company in a land rich in resources but offering only meagre markets, Massey had started to become more international than Canadian. The process of internationalization — for a company engaged in manufacturing rather than in Canadian industries like mining, forest products, or energy — was without precedent. And it also brought with it unprecedented difficulties. Massey had to rely on foreign countries, and on the good will of foreign buyers and foreign governments, for more than 90 percent of its business.

Not unnaturally, the stress was put on the creation of a highly cosmopolitan company. The senior executives have more often been American, British, or European than Canadian. In the last fifty years, the company has never had a Canadian-born chief executive. Still, Massey has had its home in Canada and has been owned continuously by Canadians.

The outcome of this curious mix has been a company that has been extraordinarily successful at times, and dismally unsuccessful at others. While Massey's successes and failures can be traced to a variety of factors — its management, the personalities that have run the company, the limitations of the Canadian market, and the vigor with which new policies and products have been implemented — the single greatest influence has been the boom-and-bust cycle of farm markets around the world.

Swings are not uncommon in any industry. Managements of great companies alternate between being innovative and vigorous, and indecisive and unimaginative. Years of good markets and solid earnings tend to lead to overconfidence and complacency. When times get tough, managements are replaced and new presidents emerge. But, for all that, the cycles that Massey has gone through have been unusually severe. The company has been taken to the brink of bankruptcy regularly, finding itself having to stave off its creditors and resort to policies of retrenchment and cutbacks. In short, Massey's history can be seen as one long battle for survival. Unable to count on any single rich captive market, such as its main competitors John Deere and International Harvester were able to build up in the U.S., Massey has been vulnerable to the fluctuations of farm economies around the world.

In the 1920s, Massey had one failure after another in its efforts to compete with the major American firms and develop new machinery and new markets. In the 1930s, it was the victim of the dust bowl and the Depression. The setbacks it encountered were only finally overcome when it switched from farm production to military hardware in the 1940s. By the middle of the 1950s, Massey was caught unprepared by another turn in world farm markets. After a power struggle that removed Duncan, a new management under Eric Phillips and Albert Thornbrough, performed another rescue. Towards the end of the 1960s, as the global economy moved from conditions of scarcity to an overabundance of grain and food supplies, Massey was in trouble again.

For a brief time, its management still under Thornbrough managed to turn things around. But fundamental forces in the world economy were working against the company. Massey had grown without regard to any prudent assessment of the way markets for its products would develop.

In the crisis of the past three years, corrective action has come late. And while the company under Victor Rice has managed to piece together a rescue plan, there is no guarantee that Massey will be able to make itself viable and survive. Under its financing plan, the company has provided itself with extra time to recoup in the belief that the worst farm market since the 1930s will not continue for much longer. The assumption is a reasonable one. But Massey's accumulated losses of the four years from 1978 to 1981 amount to over $600 million. If it is to survive under the pressure of its financial deadlines, Massey will have to make itself profitable.

Victor Rice's opinion on what will happen is an optimistic one. All the signs, from the low level of world grain stocks to the figures on farmers' buying intentions, point to a rebound. Massey has promised its lenders that it will trim its manpower to 44,000 employees; that it will keep production in line with demand; that it will find new low-cost suppliers and put more emphasis on quality control. It will soon offer a new line of rotary combine harvesters to its North American customers, and will also develop new dieselized versions of gas engines for the auto industry through Perkins, which has been given its own corporate identity.

To come up with better sales and profits, the company will focus on North America. It will attempt to raise its share of the combines market, a traditional area of strength, from just 20 percent. To do this, it will take aim at John Deere, which now has the largest share of the market. And it will promote its four-wheel-drive 4000 series tractor (Massey has under 10 percent of the large-tractor market).

According to the company's plans, a combination of growing

markets and an expansion of Massey's share of the market will take sales to $4.7 billion in 1983 and profits to $230 million. This performance will give the company the breakthrough that is needed even if markets should turn down subsequently.

Against this, a more sombre assessment has to be made. In the last few years, the single factor that has turned Massey's situation from one of difficulty to one of desperation has been the impact of inflation and rising interest rates. The crisis has been magnified both because of the impact interest costs have had on Massey's unwieldy debt and because high interest rates forced farmers to retreat from the market in 1978, and again in 1980 and 1981. While there are indications that farmers would like to buy, and that farm incomes will improve, there are few signs that interest rates will come down. Massey has been given a reprieve through the interest forgiveness plan and the restructuring of its debt, but that in itself does not provide a solution. Can there be a sufficient rise in market demand for a real recovery to occur? And will it take place soon enough?

Aside from its financial predicament, Massey has other areas of weakness. It is still a high-cost manufacturer, with plants scattered around the world. It is concentrated in Britain where high operating expenses have made its products costly. It is in an industry where there are too many U.S. and European firms carving up small markets; where margins are low in its main small-tractor business; and where it is competing against firms that have have been able to spend more on product development and research. Moreover, markets in the Third World — a strong sales area in the past — have weakened considerably and, in the next few years, Massey faces the prospect of competition there from the Japanese.

The fundamental factor determining Massey's fate will remain the same as in the past. The company will have to rely on the willingness, and capacity, of farmers around the world, but particularly in North America, to reinvest and purchase new equipment.

Over the next decade, there is no doubt that world food needs will increase substantially. And so will the prices farmers receive for their produce. The question that remains unanswered is whether costs will run ahead of prices. Will farmers ultimately find themselves worse off rather than better off? Certainly the pattern of the last few years has been for them to be worse off.

The richest farm market in the world, the U.S., provides an example. During the 1970s, food prices continued to spiral upwards — giving the illusion that farmers were the principal beneficiaries from inflation. In fact this has not been the case. The cost of everything that farmers need, from fuel to farm equipment to pesticides, has risen faster, forcing the average

American farmer to go deeper into debt. Farm liabilities
during the 1970s tripled. And real profits for farmers in 1980
(after taking account of inflation) were actually lower than they
had been in 1963. While it is true most farm assets have
increased, this can be traced to the rising price of farmland. To
realize this gain, a farmer must either sell his land or refinance
his business, using the land as collateral. But with high interest
rates, monetizing the gains in the value of farmland only adds
to costs, and farm debts.

All this suggests that farm markets are not likely to provide
the sustained and reliable growth periods that they have in the
past. Despite the evidence of the figures and statistics, and the
reality of world food shortages, there is no guarantee that
markets will become much more expansive — and helpful to
Massey — in the years ahead. In the 1980s it is going to be
difficult to accomplish a major turnaround. Certainly this has
proved to be the case so far.

During the first half of 1981 markets were expected to
improve. In fact sales plunged by 20 percent, as the company
had to fight off rumors that it would not stay in business, and
farmers worried about low prices for their grain and high
interest rates. Even after the refinancing and the addition of new
equity, Massey will have money problems. The cash squeeze
has hurt the company, affecting both its ability to carry
inventory, and its ability to go ahead with capital outlays and
investments. All these problems will be hard to overcome.
Moreover, since the company is short of time, its survival in its
present form remains still in doubt.

Victor Rice's personal vision of Massey was given to shareholders at the annual meeting in 1981. It is of a company that
"retains the potential to be a leader in its field, and a profitable
leader, with a quality of earnings that will ensure economic
expansion year in and year out."

From its beginnings, Massey has pioneered in many areas
and has made great contributions to farm mechanization. For
nearly 140 years Massey and its offspring — the Massey-
Harris Company and Massey-Ferguson Limited — have expanded
and grown, exerting an influence over farming and food
production on a global scale. Still, the company has never yet
managed to achieve what Rice envisions for it. It continues to
be engaged in a struggle to survive.

Appendix

The Men at the Top

1970 — 1976
A. A. Thornbrough, President & Chief Executive Officer

1977
John A. McDougald, Chairman of the Board
A. A. Thornbrough, President & Chief Executive Officer

1978
Conrad M. Black, Chairman of the Board
A. A. Thornbrough, Deputy Chairman of the Board & Chief Executive Officer
Victor A. Rice, President & Chief Operating Officer

1979
Conrad M. Black, Chairman of the Board & Chief Executive Officer
A. A. Thornbrough, Deputy Chairman of the Board
Victor A. Rice, President & Chief Operating Officer

1980
Victor A. Rice, Chairman of the Board, President & Chief Executive Officer.

Changes In Leadership
Members of the Board of Directors 1975 — 1980

	The Old Guard			Under the Blacks		The Post Argus Board
	1975	1976	1977	1978	1979	1980
Albert Thornbrough	■	■	■	■	■	■
The Marquess of Abergavenny	■	■	■	■	■	■
Alex Barron	■	■	■			
Henry Borden	■	■	■			
Charles Gundy	■	■	■			
John McDougald	■	■	■			
Gilbert Humphrey	■	■	■	■		
John Leitch	■	■	■	■	■	■
A. Bruce Matthews	■	■	■	■	■	■
Maxwell Meighen	■	■	■			
John Mitchell	■	■	■			
A. M. Runciman	■	■	■	■	■	■
John Staiger	■	■	■			
E. P. Taylor	■					
Trumbull Warren	■	■	■	■	■	■
Colin Webster	■	■	■	■	■	■
The Duke of Wellington	■	■	■	■	■	■
J. P. R. Wadsworth		■	■	■	■	■
H. N. R. Jackman		■	■			
F. David Radler				■	■	■
Victor A. Rice				■	■	■
Conrad M. Black				■	■	■
Ralph N. Barford				■	■	■
G. Montegu Black				■	■	
Dixon S. Chant				■	■	
Fredrik S. Eaton				■	■	
The Hon. J. Turner						■
Charles Laidley						■

Sales Statistics   Net Sales by Markets (Millions of U.S. Dollars)

	1980* % of Total	1980* Amount $	1979* $	1978* $	1977 $	1976 $	1975 $	1974 $	1973 $
North America									
Canada	7.0	219.4	217.8	180.0	196.8	222.6	193.6	143.2	102.5
United States	26.2	819.5	839.1	635.1	699.3	634.4	593.9	476.8	410.5
Total	33.2	1,038.9	1,056.9	815.1	896.1	857.0	787.5	620.0	513.0
Europe									
United Kingdom	9.5	296.9	339.2	321.0	274.4	225.4	211.6	157.5	146.8
France	7.2	227.2	199.8	182.9	189.3	166.7	171.3	142.4	137.4
Italy	6.7	210.9	155.1	122.0	136.2	119.1	89.5	59.3	54.5
West Germany	5.0	157.2	178.9	188.4	220.2	183.1	157.5	88.0	102.6
Scandinavia	3.6	113.6	84.5	104.0	102.7	90.4	86.9	56.1	45.6
Benelux	0.9	28.3	25.8	32.8	44.2	39.8	33.3	19.0	15.8
Austria	0.6	17.8	13.8	13.2	21.9	15.9	14.4	10.3	10.9
Spain	0.3	8.2	10.8	12.0	21.6	19.3	18.6	16.7	10.2
Other	1.8	56.7	44.3	46.1	42.9	30.5	30.4	17.5	16.1
Total	35.6	1,116.8	1,052.2	1,022.4	1,053.4	890.2	813.5	566.8	539.9
Latin America									
Brazil	9.8	306.0	317.9	249.6	277.1	403.6	363.1	213.3	164.5
Mexico	2.4	75.3	53.8	42.6	20.9	37.4	35.0	19.0	11.3
Argentina	1.4	43.8	43.1	32.8	109.2	72.6	51.7	51.1	29.2
Other	1.7	53.8	42.6	39.7	48.3	35.8	51.8	32.7	23.5
Total	15.3	478.9	457.4	364.7	455.5	549.4	501.6	316.1	228.5
Africa	5.2	164.5	137.0	142.9	168.4	149.6	195.5	127.0	83.6
Asia	6.5	202.4	151.0	189.1	166.8	205.9	147.5	68.3	57.5
Australasia	4.2	130.6	118.5	96.8	121.3	121.5	108.8	92.4	74.3
Total	100.0	3,132.1	2,973.0	2,631.0	2,861.5	2,773.6	2,554.4	1,790.6	1,496.8

Net Sales by Products (Millions of U.S Dollars)

Farm Machinery*	% of Total Sales	1980	1979	1978	1977	1976	1975	1974	1973
Tractors	42.0	1,315.1	1,256.6	1,107.8	1,201.5	1,186.0	1,038.2	673.3	577.0
Grain Harvesting	14.7	460.1	445.1	341.7	333.3	343.7	353.5	246.3	186.0
Hay Harvesting	1.6	50.6	46.1	49.8	60.3	55.1	51.9	42.8	41.1
Other Products	5.6	174.5	176.3	209.3	226.9	234.0	226.6	191.9	153.1
Engines									
Net Sales*	16.3	510.9	427.3	478.2	330.5	285.6	233.4	158.2	133.7
Industrial and Construction Machinery†									
Machines	5.2	164.3	195.7	170.8	338.0	314.5	285.8	203.5	181.8
Parts Total×	14.6	456.6	425.9	373.4	371.0	354.7	365.0	274.6	224.1
All Sales Total	100.0	3,132.1	2,973.0	2,631.0	2,861.5	2,773.6	2,554.4	1,790.6	1,496.8

* after deducting M–F
† no construction machinery after 1977
× for farm machinery, engines and industrial and construction machinery

1972	1971	1970	1969	1968	1967	1966	1965	1964	1963	1962	1961
$	$	$	$	$	$	$	$	$	$	$	$
77.8	69.1	65.2	79.9	66.8	84.4	89.1	82.5	75.6	70.8	58.1	50.8
336.4	293.6	247.8	285.9	256.6	263.1	268.5	208.4	174.2	154.0	134.2	132.2
414.2	362.7	313.0	365.8	323.4	347.5	357.6	290.9	249.8	224.8	192.3	183.0
128.9	116.2	114.3	106.1	95.5	95.1	95.6	83.8	88.2	84.9	72.8	73.8
119.2	95.6	88.5	110.7	93.1	89.0	95.3	85.9	85.3	74.2	66.3	60.2
45.7	39.4	41.8	37.3	30.9	28.6	25.4	22.4	19.1	18.1	13.4	7.3
62.4	56.4	57.8	47.5	35.6	28.8	43.1	42.7	43.6	34.8	36.0	33.1
42.3	41.3	39.7	34.7	29.7	33.3	38.5	34.3	35.0	29.2	35.1	30.7
10.5	9.3	11.6	8.4	7.6	6.7	10.3	8.2	7.4	5.4	6.9	6.4
8.3	10.1	8.2	6.0	5.5	6.2	7.0	5.4	4.3	4.5	5.6	4.4
8.3	4.3	8.1	9.0	4.1	5.3	4.4	3.3	3.7	4.7	5.7	2.2
14.3	13.9	14.0	11.5	10.5	8.7	9.6	8.5	9.1	5.8	6.4	8.5
439.9	386.5	384.0	371.2	312.5	301.7	329.2	294.5	295.7	261.6	248.2	226.6
121.5	76.4	58.8	43.0	38.0	20.5	24.4	19.4	28.7	19.0	9.9	5.6
11.5	11.5	11.7	11.8	10.3	11.8	9.6	7.5	6.2	4.6	3.9	2.9
15.5	10.1	9.6	3.8	2.2	2.9	1.6	2.1	2.7	1.5	1.3	1.7
15.9	21.8	14.7	11.5	10.5	7.5	9.8	9.1	11.1	8.9	8.9	7.9
164.4	119.8	94.8	70.1	61.0	42.7	45.4	38.1	48.7	34.0	24.0	18.1
77.6	76.9	68.7	68.0	58.2	54.0	43.0	44.0	41.4	39.8	28.9	19.5
43.4	40.8	33.5	39.0	33.1	41.1	33.6	29.2	25.1	30.2	22.1	21.5
50.5	42.6	43.9	55.3	59.8	57.8	53.4	51.2	53.4	43.9	35.9	33.5
1,190.0	1,029.3	937.9	969.4	848.0	844.8	862.2	747.9	714.1	634.3	551.4	502.2

1972	1971	1970	1969	1968	1967	1966	1965	1964	1963	1962	1961
462.1	396.0	331.0	339.6	317.8	335.6	359.8	294.0	293.5	275.8	243.5	219.5
149.5	128.0	99.4	148.5	135.6	140.6	145.7	131.5	118.7	89.7	77.5	81.6
26.3	29.3	26.1	30.1	26.2	27.2	32.1	32.4	32.0	31.5	27.6	25.1
112.3	97.0	99.6	94.6	83.2	87.0	80.1	71.9	60.0	53.2	46.0	45.3
117.7	107.0	121.6	104.9	84.6	78.5	76.2	74.5	77.7	66.3	58.1	43.4
147.0	121.6	128.1	128.0	88.6	73.3	69.3	54.4	49.0	41.3	33.2	26.5
175.1	150.4	132.1	123.7	112.0	102.6	99.0	89.2	83.2	76.5	65.5	60.8
1,190.0	1,029.3	937.9	969.4	848.0	844.8	862.2	747.9	714.1	634.3	551.4	502.2

Financial Statistics

(Millions of U.S. Dollars)

		1980	1979	1978*	1977*	1976*	1975*
Summary of Operations	Net sales	$	3,132	2,973	2,631	2,861	2,774	2,554
	Gross profit	$	556	573	512	603	533	511
	Net expenses (excluding interest)	$	501	377	453	417	277	272
	Interest expense (net)	$	259	164	155	151	101	99
	Provision for Reorganization Expense	$	29	95	73			
	Income tax recovery (expense)	$	10	6	(17)	(17)	(57)	(49)
	Finance subsidiaries and Associate Cos.	$	23	21	19	14	10	9
	(Loss) Profit from Continuing Operations	$	(200)	(36)	(167)	32	108	100
	Loss from Discontinued Operations	$	(25)	(23)	(95)			
	(Loss) Profit before Extraordinary Item	$	(225)	(59)	(262)	32	108	100
	Extraordinary Item			95				
	Net (loss) income	$	(225)	37	(262)	32	108	100
	Operating (loss) profit	$	(139)	30	(133)	77	126	111
	Dividends – Common	$			4	19	18	13
	– Preferred	$			2	10	7	2
	(Loss) income retained	$	(225)	37	(268)	3	83	85
Financial Condition	Working capital	$	213	426	431	703	739	643
	Additions to fixed assets	$	46	77	99	147	175	170
	Depreciation and amortization	$	80	88	77	69	54	45
	Total assets	$	2,828	2,745	2,573	2,620	2,323	2,015
	Current ratio		1.1	1.3	1.3	1.6	1.8	1.8
	Asset turnover ratio		1.1	1.1	1.0	1.1	1.2	1.3
	Debt/equity ratio**		4.6	2.1	2.1	1.2	0.9	1.0
Liabilities and Shareholders' Equity	Current	$	1,894	1,510	1,297	1,094	894	831
	Other	$	581	658	735	712	619	515
	Shareholders' equity**	$	353	578	541	813	811	669
	Return on closing equity**	%	(63.8)	6.4	(48.5)	3.9	13.3	14.9

As a Per Cent os Sales	Cost of goods sold, at average exchange rates	%	**82.0**	80.1	81.1	78.0	76.8	77.6
	Effect of foreign currency exchange rate changes	%	**0.2**	0.6	(0.6)	0.9	4.0	2.4
	Gross margin	%	**17.7**	19.3	19.5	21.1	19.2	20.0
	Marketing, general and administrative	%	**12.9**	11.8	12.6	11.9	11.5	10.6
	Engineering and product development	%	**1.9**	2.0	2.2	2.4	2.2	2.2
	(Loss) profit before Items Shown Below	%	**(6.5)**	1.1	(3.7)	1.2	5.6	5.5
	Provision for Reorganization Expense	%	**(0.9)**	3.2	2.8			
	Net (Loss) Income	%	**(7.2)**	1.2	(10.0)	1.1	3.9	3.9
	Operating (loss) profit	%	**(4.4)**	1.0	(5.0)	2.7	4.6	4.4
Per Common Share ($U.S.)	Net Sales	$	**171.62**	162.90	144.16	156.76	152.00	139.94
	(Loss) Income (after cumulative dividends on preferred shares)	$	**(12.79)**	1.58	(14.84)	1.21	5.53	5.37
	(Loss) Income retained	$	**(12.79)**	1.58	(14.68)	0.16	4.55	4.66
	Equity	$	**12.87**	25.66	24.08	39.06	38.92	34.53
	Toronto Stock Exchange quotes, High	$	**13⅝**	15½	20¼	24⅛	32	18⅛
	($Canadian) Low	$	**5¾**	9½	9½	16⅛	16¾	12⅛
	Dividends declared ($Canadian)	$			0.25	1.08	1.00	0.70
	Dividends paid ($Canadian)	$			0.25	1.08	1.00	0.90
Shareholders/ Employees	Shareholders — Common shares		**28,351**	29,926	31,353	30,619	31,039	35,844
	— Preferred shares		**9,669**	10,613	11,370	10,208	10,620	5.046
	Employees (at year end)		**41,690**	56,233	57,983	67,151	68,200	64,572
	Common shares outstanding (thousands)		**18,250**	18,250	18,250	18,250	18,250	18,250
	Preferred shares outstanding (thousands)		**3,825**	3,825	3,825	3,999	3,999	1,600

*Results for 1980, 1979, and 1978 include the construction machinery business as discontinued operations. It is not practicable to segregate construction machinery operations for years prior to 1978.

**Accounting Series Release 268 (see Note 15(d) of the SEC requires that certain preferred shares used in the calculation of equity ratios be considered as debt rather than equity. Under this assumption, the equity ratios presented above would be as follows: debt/equity, 1980—6.7, 1979—2.7, 1978—2.8, 1977—1.50, 1976—1.1, 1975—1.1, return on closing equity, 1980—(87.5)%, 1979—7.6%, 1978—(58.9)%, 1977—4.5%, 1976—15.2%, 1975—15.9%.

Prices on the Canadian Stock Exchange for Massey Common Shares 1966 — 1980

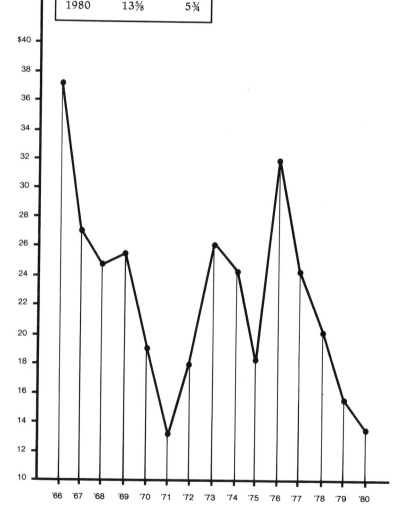

	High	Low
1966	37¼	20
1967	27¼	15¼
1968	24¾	14
1969	25½	15¾
1970	19¼	8½
1971	13	9
1972	18⅛	8½
1973	26⅛	15¾
1974	24⅜	11½
1975	18⅛	12⅛
1976	32	16¾
1977	24⅛	16⅛
1978	20¼	9½
1979	15½	9½
1980	13⅝	5¾

Dividends Paid on Massey Common Shares
1966 — 1980

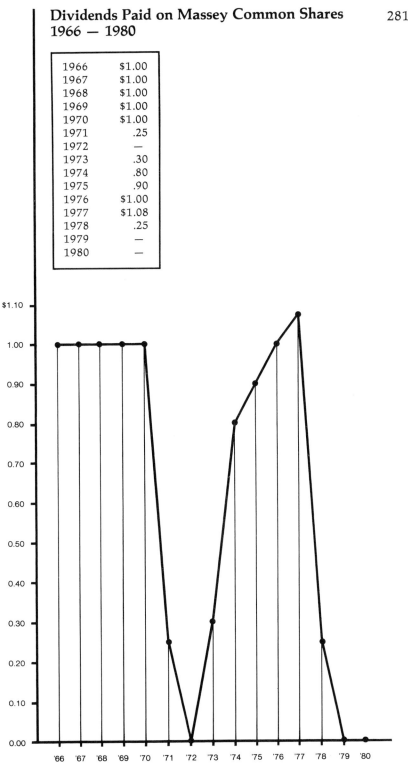

1966	$1.00
1967	$1.00
1968	$1.00
1969	$1.00
1970	$1.00
1971	.25
1972	—
1973	.30
1974	.80
1975	.90
1976	$1.00
1977	$1.08
1978	.25
1979	—
1980	—

Growth of Total Assets
1966 — 1980

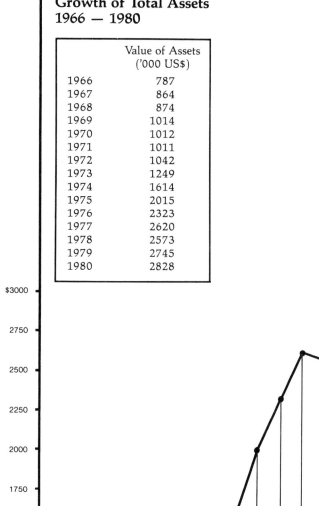

	Value of Assets ('000 US$)
1966	787
1967	864
1968	874
1969	1014
1970	1012
1971	1011
1972	1042
1973	1249
1974	1614
1975	2015
1976	2323
1977	2620
1978	2573
1979	2745
1980	2828

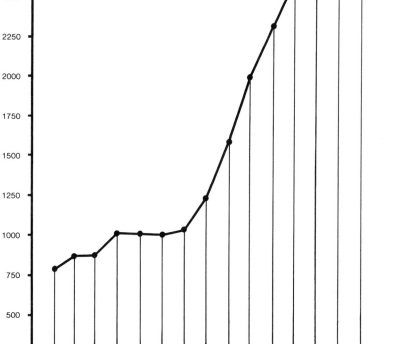

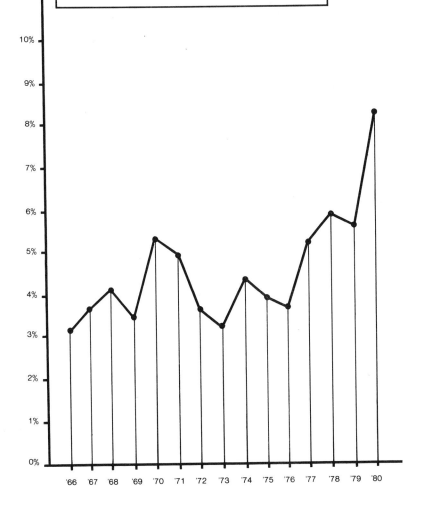

		in '000 US$	
Year	Net Sales	Net Interest	%
1966	862	27	3.13
1967	845	31	3.67
1968	848	35	4.13
1969	969	33	3.41
1970	938	50	5.33
1971	1029	51	4.96
1972	1192	43	3.61
1973	1506	48	3.19
1974	1785	78	4.37
1975	2554	99	3.88
1976	2774	101	3.64
1977	2861	151	5.28
1978	2631	155	5.89
1979	2973	164	5.52
1980	3132	259	8.27

Operating Companies — Facilities and Products

Farm and Industrial Machinery

Argentina
Massey-Ferguson Argentina S.A.
agricultural tractors.

Australia
Massey-Ferguson (Australia) Limited
sugar cane harvesters, loaders,
backhoes, combines, implements.

Brazil
Massey-Ferguson Perkins S.A.
combines, implements, backhoes,
agricultural and industrial
tractors.

Canada
Massey-Ferguson Industries Limited
Brantford Locations
combines, combine cabs, grey iron
castings, plows, mowers, rakes and
other implements, combine and
tractor components, steel stamp-
ings.
Toronto Plant
balers, corn heads, tractor cabs,
combine and tractor components.
Cambridge Foundry
grey iron and nodular castings.

France
Massey-Ferguson S.A.
agricultural tractors, tractor
components, combines, balers,
tractor cabs, components, grey
iron castings.

Italy
Massey-Ferguson S.p.A.
industrial crawler tractors: agricultural and industrial machinery
components, agricultural tractor
components, agricultural wheel
tractors.

United Kingdom
Massey-Ferguson (United Kingdom)
Limited
tractor components, agricultural
and industrial tractors, axles, gear-
boxes, other components, tractor-
backhoe-loaders and tractor
components.

United States
Massey-Ferguson Inc.
4-wheel-drive agricultural
tractors, tractor-backhoe-loaders,
agricultural and industrial
tractors, transmissions and axles,
gears and shafts.

West Germany
Massey-Ferguson GmbH
hydraulic cylinders and
components.
Gebr. Eicher GmbH
tractors, implements.

Engines

Australia
Perkins Engines Australia Pty. Ltd.
industrial diesel engine assembly,
engine reconditioning.

Brazil
Massey-Ferguson Perkins S.A.
diesel engines.
Progresso Metalfrit S.A.
grey iron castings.

France
Moteurs Perkins S.A.
diesel and gasoline engine
assembly.

United Kingdom
Perkins Engines Group Limited
diesel and gasoline engines, engine
reconditioning, engine
components.

United States
Perkins Engines, Inc.
diesel engine assembly.

Parts Facilities

The Company has 31 major parts
warehouses in 10 countries. The
largest facilities are located at:
Racine, U.S.A.
Urmston, U.K.
Brantford, Canada
Athis Mons, France
Eschwege, West Germany

Associate Companies and Per Cent Owned

Argentina
Perkins Argentina S.A.I.C. 30%
diesel engines.

Brazil
Piratininga, Implementos
Agricolas S.A. 40%
farm implements.

India
Tractors and Farm Equipment
Limited 49%
tractors, implements.

Italy
Simmel S.p.A. $33\frac{1}{3}$%
crawler tractor components.

Libya
Libyan Tractor Company 33 1/3%
tractors.

Malawi
Agrimal (Malawi) Limited 20%
hoes, animal draft equipment.

Mexico
Motores Perkins S.A. 21%
diesel engines.

Morocco
Compagnie Maghrebine de Materiels
Agricoles et Industriels. S.A. 24%
tractors.

Peru
Tractores Andinos S.A. 49%
tractors.
Motores Diesel Andinos S.A. 24%
diesel engines.

South Africa
FedMech Holdings Limited 25%
implements, tractor accessories,
attachments, industrial loaders,
transport systems, harvesting
machinery, trailers.

Other Licensee Locations

Farm and Industrial Machinery: Greece,
Iran, Japan, Kenya, Malaysia,
Mexico, Pakistan, Poland, Portugal,
Saudi Arabia, Spain, Thailand,
Turkey, Uruguay

Engines: Bulgaria, Greece, India, Iran,
Pakistan, Poland, South Africa,
South Korea, Spain, Turkey,
Uruguay, Yugoslavia.

Index

Note: 'n' following a page number indicates a footnote reference.